# Coastal Deposits

# Coastal Deposits: Environmental Implications, Mathematical Modeling and Technological Development

Editors

**Marta Pérez-Arlucea**
**Rita González-Villanueva**

MDPI • Basel • Beijing • Wuhan • Barcelona • Belgrade • Manchester • Tokyo • Cluj • Tianjin

*Editors*
Marta Pérez-Arlucea
University of Vigo
Spain

Rita González-Villanueva
University of Vigo
Spain

*Editorial Office*
MDPI
St. Alban-Anlage 66
4052 Basel, Switzerland

This is a reprint of articles from the Special Issue published online in the open access journal *Applied Sciences* (ISSN 2076-3417) (available at: https://www.mdpi.com/journal/applsci/special_issues/coastal_deposits).

For citation purposes, cite each article independently as indicated on the article page online and as indicated below:

LastName, A.A.; LastName, B.B.; LastName, C.C. Article Title. *Journal Name* **Year**, *Volume Number*, Page Range.

**ISBN 978-3-0365-1144-3 (Hbk)**
**ISBN 978-3-0365-1145-0 (PDF)**

# Contents

# About the Editors

**Marta Pérez-Arlucea** PhD in Geological Sciences from the Complutense University (Madrid, Spain). Dept. of Marine Sciences, University of Vigo. Research interests: marine geology; sedimentary processes and architecture in the coast and deep marine environments

**Rita González-Villanueva** has a PhD in Marine Science by the Vigo University (Galicia, Spain). She is a coastal researcher in the Marine Sciences Department (Vigo University), who uses GPS, GPR, core and GIS technologies to understand recent and past coastal changes. She collaborates with atmospheric modelers to understand the role of atmospheric circulation patterns, over potential extreme events that cause shifts in coastal records. She also has a broader interest in the application of GIS analyses to investigate environmental issues, such as habitat mapping, coastal erosion, and extreme events impact.

*Editorial*

# Applied Sciences: "Coastal Deposits: Environmental Implications, Mathematical Modeling and Technological Development"

**Marta Pérez-Arlucea * and Rita González-Villanueva ***

Centro de Investigación Mariña, Department of Marine Geosciences and Territorial Planification, Universidade de Vigo, XM1, 36310 Vigo, Spain
* Correspondence: marlucea@uvigo.es (M.P.-A.); ritagonzalez@uvigo.es (R.G.-V.)

**Citation:** Pérez-Arlucea, M.; González-Villanueva, R. Applied Sciences: "Coastal Deposits: Environmental Implications, Mathematical Modeling and Technological Development". *Appl. Sci.* **2021**, *11*, 119. https://dx.doi.org/10.3390/app11010119

Received: 26 November 2020
Accepted: 21 December 2020
Published: 24 December 2020

**Publisher's Note:** MDPI stays neutral with regard to jurisdictional claims in published maps and institutional affiliations.

A large percentage of the world's population lives along the coastal zones, with more than half of the world's population living in coastal areas. In such a manner, human beings depend on the coasts and oceans for their survival. In more than one way, the coastal areas remain an untapped and untamed resource; however, they are not uninfluenced by humankind. The coastal areas are affected by the urbanization, industrial progress, and the progressive expansion of a society of leisure and crowd tourism. Yet, it is an incomplete view in which an essential element we often forget: the coastal environment. Global climate change may accentuate the social pressure on the latter, namely increases temperatures and sea-level rise. On the other hand, extreme events are more frequent in recent decades.

The coastal zone is the limit between the open ocean and the continental interiors, and protect from coastal natural hazards by absorbing energy and momentum fluxes from the ocean. Examples of this are the coastal dunes, beaches, marshlands and diverse bio shields (reefs, seagrass, mangles and other types of coastal vegetation, etc.). A variety of processes, operating at a range of spatial scales, govern the stability of coastal systems. A good example of how it works and the complexity of coastal systems is presented in this Special Issue by F-Pedrera et al. [1]. The work analyzes the water circulation in a micro-estuarine environment, in the Ebro Delta in the Mediterranean coast, based on mathematical models. Results show complex dynamics of water circulation due to different processes, including wind forcing mechanisms. Another exemplification is provided by the work of Chang Y. et al. [2] in which research-based on video monitoring systems in South Korea is introduced, intending to prevent erosion by waves in a beach environment. Understanding wave incidence and maritime climate using remote sensing techniques improve the advancement in prediction models of sediment accumulation and erosion cycles. Both works highlight the importance of understanding the natural processes for the correct management and coastal protection.

These systems are highly vulnerable to storm-induced hazards, namely coastal erosion and flooding. Combine these with the effect of human activities, which have led to changes in these processes at all spatial scales, resulting in a severe threat for the survival of coastal ecosystems. Indeed, two works in this Special Issue cope with the human-derived impacts. On one side, the work of Dao, M. et al. [3] tackles with conflicting activities within the coast in Vietnam, particularly the Titan mining industry and the use of alternative natural resources and ecosystem preservation. The method used in the evaluation is a combination of Fuzzy AHP and Fuzzy TOPSIS, two analytical techniques to assess conflicting activities. On the other, the work of Quang Tri D. et al. [4] also assesses environmental impacts due to activities at sea, in a case in point, the dredging and dumping labours, which are a source of pollutants to the marine environment. They approached this problem using meteorological data and mathematical models for pollutant dispersion. Pollution by the dispersion of particles derived from dredging spoils and other sources is another significant cause of concern regarding the impact of human activities on the marine environment.

Both pieces of research had stressed how the functionality of the coastal environments could be endangered by one single socio-economic sector, preventing the development of other sources of economy.

Moreover, considering that sea-level rise and changes in storminess are linked to climate change scenarios, the increase in risks in these highly populated areas are also expected to rise. In this context, two solutions to hinder coastal erosion are introduced in this Special Issue. First, Imran et al. [5] proposed a method via processes of biocementation, a technique consisting of inducing carbonate precipitation using microbes. Experimental results in Korea show differences in the efficiency of different species and the influence of environmental parameters. This research illustrates the development of ecofriendly techniques to protect the coast. In this vein, Nayantara el al. [6] present a research work based on carbonate precipitation using native bacteria to stabilize sandbanks in nearshore areas and minimize beach erosion.

Apart from the climate change-related factors, increase pressure on the coast leads to water pollution, overexploitation of water resources and coastal modifications (i.e., infrastructures, protection constructions and others). The related impacts put at risk coastal and marine ecosystems, which in many instances have protection figures.

In consequence, the understanding of the effects of these hazards and the impacts of humankind is crucial to prevent further environmental degradation and for increasing the safety and reducing the impacts over the exposed population in coastal areas.

These threats have made that the coastline monitoring programs and sustainable coastal management area are a priority. Ergo, information about the coastal state and dynamics are essential for coastal managers. Indeed, the research that places emphasis on the fight and the adaptation to climate change and the conservation of the oceans, seas and marine resources are priority objectives in many countries and communities. The UNESCO developed the Marine and Spatial Planning Programme (MSP), with the participation of many nations and territories, to achieve sustainable use of the marine and coastal environment and to develop a "Blue Economy" and biodiversity conservation. The European Community Union adopted two instruments for the coastal and marine protection, The Marine Strategy Framework Directive and the Integrated Coastal Zone Management, which aimed to reach a Good Environmental Status (GES 2020). There are many more strategic plans all over the world to protect the coastal and marine environments, because of the global deterioration suffered in the last decades. The goal of these programs is developing strategies for the environment protection in the frame of sustainable development, including stakeholders, managers, policymakers, private sectors and users.

In the shoot, researchers' different programs required that the scientific community use the high temporal and spatial resolution and accurate databases with coastline information to deliver to the coastal managers and policymakers to be used in the decision-making process. Yet, despite the need to monitor and manage sandy coasts, the high temporal resolution data-scarcity based on in situ measurements are limited to local areas around the world, even when these data are crucial to understanding past behaviour and calibrate predictive shoreline evolution models.

Traditional techniques for determining coastline changes include aerial photography and ground survey techniques. These conventional techniques used for coastline monitoring are expensive and time-consuming and require trained staff, which makes it challenging to have a time-updated database. Recently, advances in remote sensing, geographical information systems and mathematical modelling and techniques are essential to support conventional methods for monitoring coastline change. Additionally, free access to data from Earth Observation Satellites provides the capability to monitor the coastline in a low cost-effective manner. Furthermore, many advanced methodologies using digital images for change detection are of everyday use. In the present Special Issue, the new remote-sensing-derived index (RISI) is suggested by Fang H. et al. [7] to study resistant surface distributions from Landsat 8 OLI imagery. The aim is to evaluate the level of urbanization and growing of urban areas and the consequent changes in environmental quality in rapidly developing countries, using China as an example. The novelty of this index is that it allows discrimination of urban modified soils from bare soil, which allows the evaluation of coastal changes due to the growing urban development. We must not forget the growth of the open-source tools, which make them accessible to just about anyone. Considering all the advantages and potential of the data and the emergent methodologies availability, it is not surprising the increased interest in the scientific community to develop new methodologies to maximize the benefits of these data and platforms, notably in the maritime and coast field.

This special volume presents some of the many scientific challenges associated with nowadays coastal issues. In so doing, it highlights the need for a better understanding of the human impact over the coastal environment, and how to solve the adverse effects derived from these impacts, which will be required to both protect and make more resilient the invaluable coastal regions.

**Funding:** This research received no external funding.

**Acknowledgments:** This Special Issue is a result of a long-term work by all the authors, the reviewers, and the editorial team.

**Conflicts of Interest:** The authors declare no conflict of interest.

## References

1. Pedrera, F.; Balsells, M.; Grifoll, M.; Espino, M.; Cerralbo, P.; Sánchez-Arcilla, A. Wind-Driven Hydrodynamics in the Shallow, Micro-Tidal Estuary at the Fangar Bay (Ebro Delta, NW Mediterranean Sea). *Appl. Sci.* **2020**, *10*, 6952. [CrossRef]
2. Chang, Y.; Jin, J.; Jeong, W.; Kim, C.; Do, J. Video Monitoring of Shoreline Positions in Hujeong Beach, Korea. *Appl. Sci.* **2019**, *9*, 4984. [CrossRef]
3. Dao, M.; Nguyen, A.; Nguyen, T.; Pham, H.; Nguyen, D.; Tran, Q.; Dao, H.; Nguyen, D.; Dang, H.; Hens, L. A Hybrid Approach Using Fuzzy AHP-TOPSIS Assessing Environmental Conflicts in the Titan Mining Industry along Central Coast Vietnam. *Appl. Sci.* **2019**, *9*, 2930. [CrossRef]
4. Quang Tri, D.; Kandasamy, J.; Cao Don, N. Quantitative Assessment of the Environmental Impacts of Dredging and Dumping Activities at Sea. *Appl. Sci.* **2019**, *9*, 1703. [CrossRef]
5. Imran, M.; Kimura, S.; Nakashima, K.; Evelpidou, N.; Kawasaki, S. Feasibility Study of Native Ureolytic Bacteria for Biocementation Towards Coastal Erosion Protection by MICP Method. *Appl. Sci.* **2019**, *9*, 4462. [CrossRef]
6. Nayanthara, P.; Dassanayake, A.; Nakashima, K.; Kawasaki, S. Microbial Induced Carbonate Precipitation Using a Native Inland Bacterium for Beach Sand Stabilization in Nearshore Areas. *Appl. Sci.* **2019**, *9*, 3201. [CrossRef]
7. Fang, H.; Wei, Y.; Dai, Q. A Novel Remote Sensing Index for Extracting Impervious Surface Distribution from Landsat 8 OLI Imagery. *Appl. Sci.* **2019**, *9*, 2631. [CrossRef]

 *applied sciences*

*Article*

# Quantitative Assessment of the Environmental Impacts of Dredging and Dumping Activities at Sea

**Doan Quang Tri** [1,*]**, Jaya Kandasamy** [2] **and Nguyen Cao Don** [3]

[1] Sustainable Management of Natural Resources and Environment Research Group, Faculty of Environment and Labour Safety, Ton Duc Thang University, Ho Chi Minh 70000, Vietnam
[2] School of Civil and Environmental Engineering, Faculty of Engineering and Information Technology CB11.11.213, University of Technology, Sydney 2007, Australia; jaya.kandasamy@uts.edu.au
[3] Water Resources Institute, No. 8 Phao Dai Lang str., Dong Da, Hanoi 10000, Vietnam; ncaodonwru@gmail.com
* Correspondence: doanquangtri@tdtu.edu.vn; Tel.: +84-98-892-8471

Received: 15 March 2019; Accepted: 19 April 2019; Published: 25 April 2019

**Abstract:** The dumping of dredge materials often raises concerns about the release of pollutants to the marine environment. Wind data from the Global Forecast System (GFS) model was used to simulate the wind-wave propagation from offshore in a two-dimensional (2D) model during September and October 2016. The calibration and validation of the 2D model showed a high conformity in both the phases and amplitude between the observed and simulated data. The 2D mud transport simulation results of three scenarios showed that the concentration of suspended material in the third scenario tested (scenario 3) was greater than 0.004 kg/m$^3$ in the low tide, spreading to a 9 km$^2$ area, and in the high tide, the concentration was 0.004 kg/m$^3$ in a 6 km$^2$ area. Finally, the results of 2D particle tracking (PT) showed changes in the seabed due to the concentration of dredged material, and its dump (approximately 180 days) increased from 0.08 m to 0.16 m in 2.85 ha. In scenario 3, the element block moved quite far—approximately 2.9 km—from the dredge position. Therefore, the simulation results were qualified, as the dredging position situated far from the sea is significantly affected by the direction and velocity of wave-wind in the dredging position.

**Keywords:** Quy Nhon port; dredged materials; dumping location; mud transport; particle tracking; 2D model

---

## 1. Introduction

Dredging is a crucial activity, regularly carried out to maintain safe passage for boats and ships at ports and harbors [1–6]. In another sense, dredging is important for maintaining harbor operations, i.e., importing and exporting goods. However, the disposal of dredged spoil at sea can potentially affect the marine environment and its ecosystems [7,8]. The dredging and dumping of spoils at sea inevitably increase the turbidity levels and settlement of fine sediment over an extended area [9,10]. This can propagate to a large area around the dumped location [11–18]. The dumping of spoils can lead to the smothering of the seabed environment, coral reefs, egg-laying nests of fish and can disturb their navigations [19–21]. Finally, dredging and dumping change the topography of the seabed at both the dredging and dumping locations and therefore, can change local flow patterns [22]. As a result, dredging and its related activities should be monitored, controlled and their impacts on the marine environments should be evaluated [23].

The Qui Nhon Port is one of the largest international ports in Central Viet Nam, with a capability to berth ships up to 30,000 tons at the regular operating frequency [24]. The port serves for the exchange of goods between Binh Dinh Province and the Central and Central Highland Viet Nam, as well as for Cambodia, Laos, and Thailand. To maintain navigation, dredging is usually carried out twice per year. Existing considerations are the method of dredging and location of suitable locations to dump the spoils without adversely affecting the surrounding environment. Diggers are used to dredge the port area. The dredged spoils are transported in hopper barges that open at the bottom. The barges have specialized designs and are typically used for transporting and disposing of dredged materials, in order to ensure environmental protection in a cost-effective manner. In particular, this method can minimize the diffusion of dredged materials and limit the increased turbidity in seawater. In this manner, it can minimize the impacts on the surrounding ecological environment.

To evaluate the effect of dredging and dumping activities on the marine environment, previous studies have mostly used two- and three-dimensional models [25–32] based on field data, calibration and validation for realistic assessment of the physical process and simulation of the sediment dispersion of the bed load movement of the dredged materials near the port [33–41]. This study combines the two-dimensional (2D) numerical approach (using MIKE 21) with numerical weather predictions (Global Forecast System (GFS) model) and is a new modeling approach for the assessment of dredging and dumping activities in Viet Nam. The study develops a two-dimensional hydrodynamic advection-diffusion model to simulate the propagation of contaminants in dredged materials to the sea with the assumption that the total volume of dredged materials from Quy Nhon port is transported by barges to the dumping location. The principal objective was to assess the impacts of dredging and dumping using a hybrid modeling approach (numerical weather prediction and hydrodynamics). The aims of this study were (1) the simulation of wind-wave propagation from offshore to the study area using wind data from the GFS global model in a 2D model, (2) the validation and calibration of wave height in a 2D model, (3) the validation and calibration of the hydraulics using the 2D model, (4) the simulation of the sediment transport process in a 2D mud transport (MT) model, and (5) the simulation of the diffusion of materials at the dumping location in a 2D particle tracking (PT) model.

## 2. Materials and Methods

### 2.1. Description of Study Site

The study site is located inside Quy Nhon and the Thi Nai Port, encompassing an area of 95.1 ha (Figure 1). Quy Nhon Port, 68.10 ha in surface area, belongs to the Hai Cang commune. Its eastern boundary is adjacent to vessels operating at the port, Dong Da Port is to the west, boat shelters are located at the Thi Nai lagoon in the south. Quy Nhon Port is one of the ten largest ports in Viet Nam and serves vessels that can load up to 30,000 tons with the regular operating frequency, and the port capacity is 10 million tons per year. The dredging and maintenance requirements of Quy Nhon Port play an important role in the transportation, evaluation, and selection of dredging location and dredging volume. The dredging area is 7.6 km from the port, 4.6 km from Hai Giang seashore, and 8.7 km from the Ghenh Rang tourist attraction (Figure 1). The water depth at the dredging position is approximately −28 m to −30 m.

A dredger boat was used to dredge the bed surrounding the port. The spoils are transported by means of a hopper barge with a bottom discharge. The spoils can be dumped at sea by opening the floor of the barge, minimizing the diffusion of these materials in water and the turbidity of the ocean, thereby reducing the negative effects on the surrounding ecological environment. The impacts of discharge on the surrounding environment were evaluated using a combination of spectral wind-wave and 2D flow model, the mud transport module and the particle tracking module.

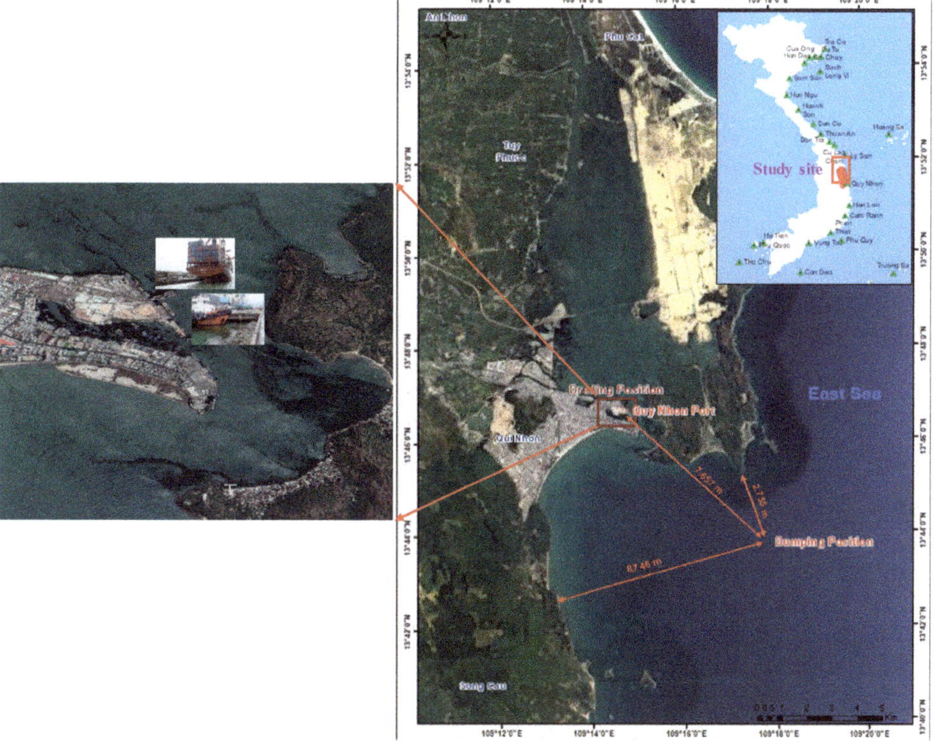

**Figure 1.** Location of study area (Quy Nhon Port, 2012).

## 2.2. Model Description

In this study, the wave field data was an important factor to support the calculated water quality spread process. A two-dimensional (2D) spectral wind-wave (SW) model was used to simulate the wave field in the study area. This model was used to calculate the wave field based on unstructured meshes. This model calculated the growth, decline and transmitted waves generated by wind and swells in offshore and nearshore areas. The dynamics of the gravity wave was simulated based on the equation of the wave action density [42,43]. The formula was used in the Cartesian coordinate system for small areas; spherical polar coordinates were applied to large areas. The conservation equation for wave action is expressed as follows:

$$\frac{\partial N}{\partial t} + \nabla \cdot (\overline{v} N) = \frac{S}{\sigma} \tag{1}$$

where $N(\overline{x}, \sigma, \theta, t)$ is the action density; $t$ is the time; $\overline{x} = (x, y)$ is the Cartesian coordinates; $\overline{v} = \left(c_x, c_y, c_\sigma, c_\theta\right)$ is the propagation velocity of a wave group in the four-dimensional phase space $\overline{x}$, $\sigma$ and $\theta$; $S$ is the source term for the energy balance equation, and $\nabla$ is the four-dimensional differential operator in the $\overline{x}$, $\sigma, \theta$-space.

The governing equations of unsteady 2D flow are based on the nonlinear, vertically integrated 2D equations of conservation of mass and momentum to delineate variations in the flow and water level of all the grids [44]. The continuity equation representing the conservation of mass is given as follows:

$$\frac{\partial \varsigma}{\partial t} + \frac{\partial p}{\partial x} + \frac{\partial q}{\partial y} = 0 \tag{2}$$

The momentum equation in the x-direction is as follows:

$$\frac{\partial p}{\partial t} + \frac{\partial}{\partial x}\left(\frac{p^2}{h}\right) + \frac{\partial}{\partial y}\left(\frac{pq}{h}\right) + gh\frac{\partial \varsigma}{\partial x} + \frac{gp\sqrt{p^2+q^2}}{C^2 \cdot h^2} - \frac{1}{\rho_w}\left[\frac{\partial}{\partial x}(h\tau_{xx}) + \frac{\partial}{\partial y}(h\tau_{yy})\right] - \Omega_q - fVV_x + \frac{h}{\rho_w}\frac{\partial}{\partial x}(p_a) = 0 \tag{3}$$

The momentum equation in the y-direction is as follows:

$$\frac{\partial q}{\partial t} + \frac{\partial}{\partial y}\left(\frac{q^2}{h}\right) + \frac{\partial}{\partial x}\left(\frac{pq}{h}\right) + gh\frac{\partial \varsigma}{\partial y} + \frac{gp\sqrt{p^2+q^2}}{C^2 \cdot h^2} - \frac{1}{\rho_w}\left[\frac{\partial}{\partial y}(h\tau_{yy}) + \frac{\partial}{\partial x}(h\tau_{xy})\right] - \Omega_q - fVV_y + \frac{h}{\rho_w}\frac{\partial}{\partial xy}(p_a) = 0 \tag{4}$$

where $p$ and $q$ (m³/s/m) are the fluxes in the x- and y-directions, respectively; $t$ (s) is the time, $x$ and $y$ (m) are the Cartesian coordinates; $h$ (m) is the water depth; $d$ is the time-varying water depth (m); $g$ (9.81 m/s²) is the acceleration due to gravity; $\varsigma$ (m) is the sea surface elevation; $C$ is a Chezy resistance parameter (m$^{1/2}$/s); $f(V)$ is the wind friction factor; $V$, $V_x$, and $V_y$ are the wind speed and its components in the x- and y-directions (m/s), respectively; $\Omega$ is the Coriolis parameter, which is latitude dependent(s$^{-1}$); $Pa$ is the atmospheric pressure (kg/m/s²); $\rho_w$ is the density of water (kg/m³); and $\tau_{xx}$, $\tau_{xy}$, $\tau_{yy}$ are the components of effective shear stress.

Sediment transport formulations are built into advection-dispersion module, 2D-AD. 2D solves the advection-dispersion equation as follows:

$$\frac{\partial \bar{c}}{\partial t} + u\frac{\partial \bar{c}}{\partial x} + v\frac{\partial \bar{c}}{\partial y} = \frac{1}{h}\frac{\partial}{\partial x}\left(hD_x\frac{\partial \bar{c}}{\partial x}\right) + \frac{1}{h}\frac{\partial}{\partial y}\left(hD_y\frac{\partial \bar{c}}{\partial y}\right) + Q_L C_L\frac{1}{h} - S \tag{5}$$

where $\bar{c}$ is depth averaged mass concentration (kg·m$^{-3}$) $u$; $v$ is depth averaged flow velocities (m·s$^{-1}$); $D_x$ and $D_y$ are dispersion coefficients; $h$ is water depth; $S$ is the deposition/erosion term (km·m$^{-3}$·s$^{-1}$); $Q_L$ is the source discharge per unit horizontal area (m³·s$^{-1}$·m$^{-2}$); and $C_L$ is the concentration of the source discharge (kg·m$^{-3}$). The bottom shear stress $\tau_b$ (N·m$^{-2}$) is calculated with respect to current and waves using following equation:

$$\tau_b = \frac{1}{2}\rho f_w\left(U_b{}^2 + U_\delta{}^2 + 2U_b U_\delta \cos\beta\right) \tag{6}$$

where $\rho$ is the density of water (kg·m$^{-3}$); $f_w$ is the wave friction factor; $U_b$ is the horizontal mean wave orbital velocity at the bed (m·s$^{-1}$); $U_\delta$ is the current velocity at top of water boundary layer (m·s$^{-1}$); and $\beta$ is the angle between the average wave direction and the observed wave direction (degree). Formula for sediment deposition was originally proposed by Krone (1962) [45] as

$$S_D = w_s C_b p_d \tag{7}$$

where $S_D$ is deposition rate (kg·m$^{-3}$·s$^{-1}$); $w_s$ is settling velocity (m·s$^{-1}$); $C_b$ is near bed concentration (kg·m$^{-3}$); $p_d$ is probability of deposition which is calculated by

$$p_d = 1 - \frac{\tau_b}{\tau_{cd}}, \tau_b \leq \tau_{cd} \tag{8}$$

where $\tau_b$ is bed shear stress (N·m$^{-2}$) and $\tau_{cd}$ is critical bed shear stress for deposition (N·m$^{-2}$). Settling velocity, $w_s$, described flocculation process. Flocculation is when the concentration of sediment is high enough for the sediment flocs to influence each other settling velocity. The modification of settling velocity also computed due to salinity variation.

$$w_{s,s} = w_s\left(1 - C_1 e^{s \cdot C_2}\right) \tag{9}$$

where $w_{s,s}$ is settling velocity due to salinity variation; $C_1$ and $C_2$ are calibration parameters; and S is salinity. Erosion formula was described as soft partly consolidated sediment [46].

$$S_E = E exp\left[\alpha(\tau_b - \tau_{ce})^{1/2}\right], \tau_b > \tau_{ce} \tag{10}$$

where $S_E$ is the erosion rate (g·m$^{-2}$); $E$ is the erosion coefficient (g·m$^{-2}$·s$^{-1}$); $\tau_{ce}$ is critical bed shear stress for erosion (N·m$^{-2}$); and $\alpha$ is coefficient (m·N$^{-0.5}$).

The overall transport of particles during a time interval, $\Delta t$, results from an advective component and a dispersive component, which accounts for the non-resolved flow processes. The particle transport equation at the $i$'th time step can be expressed as:

$$\overline{x}_{i+1} = \overline{x}_i + \overline{v}·\Delta t + \underline{\underline{D}}·\overline{V} + \overline{\gamma} \tag{11}$$

where

$$\overline{V} = \begin{pmatrix} u_x \\ u_y \\ -V_{sett} \end{pmatrix}; |\overline{U}| = \sqrt{u_x^2 + u_y^2}; \underline{\underline{D}} = \frac{1}{|\overline{U}|}\begin{pmatrix} \Delta D_L & -\Delta D_T & 0 \\ \Delta D_L & \Delta D_T & 0 \\ 0 & 0 & 0 \end{pmatrix}; \overline{\gamma} = \begin{pmatrix} \Delta D_0 \\ \Delta D_0 \\ \Delta D_{0W} \end{pmatrix} \tag{12}$$

$\Delta D_L$ is longitudinal dispersion caused by turbulence; $\Delta D_T$ is transversal dispersion; $\Delta D_0$ is neutral dispersion; $\Delta D_{0W}$ is dispersion caused by wind acting on the surface; $\overline{x}_i$ is particle coordinates in three dimensions at time steps i (m); $\Delta t$ is time steps (s); $u_x$ and $u_y$ are horizontal current velocities (m/s); $V_{sett}$ is settling velocity (m/s); $D_L$ is the longitudinal dispersion coefficient (m$^2$/s); $D_T$ is the transversal dispersion coefficient (m$^2$/s); $D_0$ is the neutral dispersion coefficient (m$^2$/s); and $D_w$ is dispersion due to wind (m$^2$/s).

*2.3. Data Collection*

2.3.1. Wind Data Collection

During the period from September to November, cold weather occurs accompanied by storms, tropical cyclones and the intertropical convergence zone in the south-central region. The prevailing wind from September to November blows from the northeast with intensities that increase from 21.6% to 51.6% (the greatest intensity is in November with the value of 51.6%) (Table 1, Figure 2). The winter monsoon lasts from December to January of the following year. This is advantageous as it is a high-pressure system. Subsidiary high-pressure systems along the eastern sea of China, with frequencies ranging from 60–70%, stabilize the winter monsoon. During this period, the prevailing wind blows from the northeast, with frequencies between 44–51%. A drop in the winter monsoon is usually observed from February to March. Although the winter is still harsh, the frequency of the winter monsoon reduces to approximately 10%. The high-pressure systems along the eastern sea of China with higher intensities account for 70% of the winter monsoon, while the Pacific high-pressure systems account for the rest of the winter monsoon. The prevailing wind in Quy Nhon still blows from the northeast, with a reduced frequency between 30–40%.

**Table 1.** The frequency and "calm wind" (%) based on directions at Quy Nhon station.

| Month | Calm Wind | Direction | | | | | | | |
|---|---|---|---|---|---|---|---|---|---|
| | | North | North-East | East | South-East | South | South-West | West | North-West |
| 1 | 9.5 | 12.6 | 43.7 | 2.3 | 4.1 | 2 | 31.2 | 4 | |
| 2 | 14.7 | 8.9 | 41.2 | 3.5 | 10.3 | 6.1 | 26.6 | 3 | 0.1 |
| 3 | 19.6 | 6.7 | 30.2 | 6.1 | 26.2 | 11.7 | 16.2 | 2.5 | 0.4 |
| 4 | 23.6 | 5.9 | 18.1 | 9.1 | 36.7 | 19.4 | 9.1 | 1.1 | 0.6 |
| 5 | 26.1 | 5.3 | 13 | 7.5 | 32.3 | 21.4 | 11 | 7 | 2 |
| 6 | 25.3 | 2.8 | 6.5 | 6 | 25.8 | 17.4 | 16.8 | 21 | 3.6 |
| 7 | 22.6 | 2.4 | 5.9 | 2.6 | 23.9 | 12.3 | 18.1 | 31 | 3.6 |
| 8 | 22.8 | 2.3 | 7.1 | 3.3 | 19.2 | 9.5 | 19.6 | 34 | 4.2 |
| 9 | 25.6 | 5.7 | 21.6 | 4.5 | 17.1 | 11.6 | 21.4 | | 3.2 |
| 10 | 14.7 | 12.8 | 36.4 | 4.8 | 5.8 | 4 | 29.4 | | 0.5 |
| 11 | 7.9 | 13.1 | 51.6 | 2.5 | 1.5 | 1.1 | 26.3 | | 0.1 |
| 12 | 6.4 | 10.6 | 51 | 1.6 | 1.2 | 0.9 | 30.8 | | |

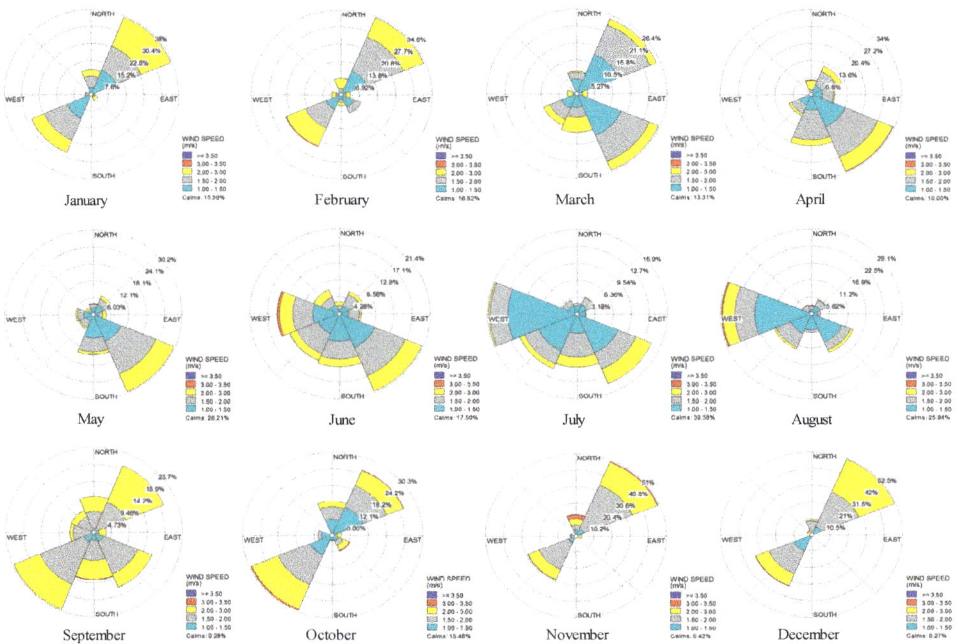


**Figure 2.** The wind speed frequency distribution and the "wind calm" (%) based on directions.

In addition to the changes in air from the polar region, high pressure cold continental winds move to the east, splitting from the Siberian high, and another subsidiary high-pressure system appears near the tropics and diffuses along the eastern sea of China. This high pressure affects the humidity, is warmer than the continental polar air mass and causes an increase in intensity of 26%, together with the southeast wind in March. Throughout the second half of April and early May, this period experiences a conflict between western and eastern systems, because cold high-pressure decreases, western low pressure moves to the east and becomes more intensive, reversing the subtropical high pressure to the east. The wind in the East Sea gradually changes direction from southeasterly to northerly, convective clouds develop strongly in mountainous areas, subtropical high pressure affects the East Sea, but the effect slowly moves to the north along with the low range equator. In this period, the prevailing wind at Quy Nhon station is in the southeast direction, with the intensity ranging from 32–37%.

From June to August, in the low range, low pressure warm air, originating from the Burmese–Indian border, tends to change direction from the east to the southeast. The monsoon season in northern

China, accompanied by tropical cyclones during the monsoon of South Asia, is closely associated to the southwest monsoon season. In the upper atmosphere, the Pacific subtropical high is degraded by the weakness and retreats to the south of the subtropical west wind. The subtropical high in the Southern Hemisphere gently ascends to the north to provide a chance for the southeast wind of the high-pressure area of Australia to move above the equator. A low equatorial ridge also ascends to the north to become active at 16° N–20° N, and accordingly, a tropical cyclone is formed to stimulate a southwest monsoon. However, due the effect of the topography, the prevailing wind at Quy Nhon Port is in the southeast direction, with a frequency from 19–26%. Moreover, during this period, the southwest and north winds dominate at Quy Nhon Port, with a frequency from 10–20%. In reality, the monsoon does not blow continuously and constantly, but periodically. Therefore, during the monsoon periods, the wind magnitude is relatively small, and the observed wind speed is 0 m/s; this is known as "calm wind" (written as "Calm" on the wind rose) (Figure 2). According to the statistics for Quy Nhon Port, the greatest frequency of calm wind was seen from April to October, and the frequency fluctuated from 23–26% (Table 1).

Another important factor of the feature of the wind regime is the wind speed. The wind speed in coastal areas is likely to be more intensive than on the Mainland. In Quy Nhon, the annual average wind speed in the mainland fluctuates from 2.4–2.6 m/s, with monthly fluctuation from 1.8–4.2 m/s (Table 2). The greatest fluctuation occurs in September and December (according to statistics over the past 5 years) and is approximately 3.1–3.2 m/s.

**Table 2.** The monthly average wind speed at Quy Nhon station from 2011–2016 (m/s).

| Month / Year | 1 | 2 | 3 | 4 | 5 | 6 | 7 | 8 | 9 | 10 | 11 | 12 |
|---|---|---|---|---|---|---|---|---|---|---|---|---|
| 2011 | 2.4 | 2.2 | 1.9 | 2.2 | 1.9 | 2.3 | 1.9 | 3.1 | 2.2 | 1.9 | 2.4 | 3.7 |
| 2012 | 2.0 | 2.1 | 2.1 | 2.0 | 2.0 | 2.0 | 2.3 | 2.8 | 4.2 | 2.5 | 2.0 | 2.8 |
| 2013 | 2.9 | 3.0 | 2.0 | 2.1 | 2.1 | 2.1 | 2.4 | 3.5 | 4.0 | 2.2 | 1.9 | 2.9 |
| 2014 | 2.5 | 2.7 | 1.9 | 2.7 | 1.8 | 2.7 | 2.2 | 2.6 | 2.7 | 1.8 | 2.5 | 2.5 |
| 2015 | 3.2 | 2.9 | 3.1 | 3.0 | 2.8 | 2.4 | 2.4 | 2.5 | 2.6 | 3.1 | 3.2 | 3.3 |
| 2016 | 2.5 | 2.7 | 2.6 | 3.0 | 1.9 | 2.2 | 2.1 | 2.2 | 3.4 | 2.7 | 3.4 | 3.3 |

### 2.3.2. The Regime of Tide

The research area is affected by two types of tide: Semidiurnal tide and irregular tide. The number of days when semidiurnal tide occurs is between 18–22 days each month, and irregular tide occurs the rest of the month. There is a small waterflow in the low tide. The time of the high tide is often longer than the ebb tide. The amplitude of tide ranges from 1.5–2 m, and this hardly changes in high tide. However, between the high tide and low tide, the magnitude changes considerably. At low tide, the flood rises and ebbs in a range of 0.5 m. At Quy Nhon station, there are four days of semidiurnal tide. The peak tide rises at 5.30 pm and ebbs most rapidly at 11.30 am in midiurnal tide, while the peak tide rises at 6.30 am and ebbs most rapidly at 7.00 am in the semidiurnal tide, with 1.53 m of the high tide and 0.61 m of the low tide.

According to statistics of Quy Nhon station, which is located at 13°46′ North–109°13′ East, the regime of tide is mostly irregular. Apart from general features, the amplitude of the tide in this area also depends on the hydraulic discharge of Kon River and Ha Thanh River that discharges to sea. The average tidal amplitude over the years at Quy Nhon ranges from 2.2 m to 2.3 m, and the maximum tidal amplitude is 2.6 m. The tidal peaks are often observed in October, November and December and the historical record was measured on 3 December 1986 at 3.06 m. The lowest peak is observed in July and August, and 0.36 m was the lowest measured tide on 8 August 1987.

2.3.3. The Characteristics of Streamflow

The characteristics of streamflow at Quy Nhon area is influenced by two monsoon periods including the northeast and southwest: (1) During the northeast monsoon, the streamflow in the estuary is virtually in the south-southwest direction. The streamflow in bottom layer is in the southeast direction going away from the shore. The streamflow speed in coastal areas is about 10–14 cm/s; the flow in the estuary area has a higher speed of about 20–25 cm/s. The flow speed at the middle layer is between 10–15 cm/s and the flow speed at the bottom layer is between 7–9 cm/s. (2) During the southwest monsoon, the flow at surface layer at the estuary area is to the north. The flow in coastal area in the south area is to the northwest. The flow at the bottom layer throughout the study area is virtually in the northeast direction from offshore to coastal. The flow speed in the northwest is between 3–14 cm/s; the flow speed in the remaining areas is between 20–28 cm/s. The flow speed at the middle layer and the bottom layer ranged between 13–18 cm/s, 5–9 cm/s, respectively. Finally, the flow at Quy Nhon area is strongly influenced by the China East Sea circulation system under the influence of monsoons.

## 3. Results

### 3.1. Calibration and Validation Spectral Wind-Wave (SW) Model

3.1.1. Establishing Mesh in SW 2D Model

To evaluate wave propagation in the study area, the bathymetry within the wave mesh was surveyed and expanded to a wide area. The computation grid is an unstructured grid of the whole study area. The computation mesh had a total of 9083 grids and 4857 nodes. The computation mesh can produce detailed results in the study area with its small and smooth grids while reducing computation steps with the use of more sparse grids offshore. An unstructured grid was established for the offshore zone with the maximum mesh size 3000 m and for the coastal area with the minimum mesh size 50 m (Figure 3b). It was then used to create the wave mesh for the study area (Figure 3a). Based on topographic features of the study area, the wave in the South China Sea hardly propagates in the Quy Nhon port. The wave at Quy Nhon port is normally generated from the wind. As the wind speed and velocity are relatively small, the wave in the port is small as well. The front of the port (the east of the port) is hardly affected by waves. Parameters were chosen based on the interaction between outside and inside the study area to outline its border (Figure 3b). The northern boundary of the study area ranges from 14°00′ N, 109°15′ E to 14°00′ N, 109°30′ E, the southern boundary ranges from 13°30′ N, 109°30′ E to 13°30′ N, 109°15′ E, and the eastern boundary ranges from 14°00′ N, 109°30′ E to 13°30′ N, 109°30′ E (Figure 3b).

Nowadays, the application of the products of meteorological models such as Global Forecast System (GFS), Weather Research and Forecasting (WRF) as input for hydrological and hydraulic 2D models is increasingly popular and effective [40,47,48]. The wind statistics are produced from the NOAA GFS model, with a resolution of 0.5 degrees and a time span of 3 h at ftp://nomads.ncdc.noaa.gov/GFS/Grid4. The Climate Data Operators (CDO) tool and Grid Analysis and Display System (GraDs) were used to produce wind direction and velocity to provide input data for a simulation spectral wind-wave model (SW). The simulation results were produced from 1 September 2017 to 31 October 2017.

The collected wind data at Ly Son station was used to evaluate the simulation results (wind direction and velocity) produced from the numerical weather prediction at the study area. The wind statistics from the GFS model are highly reliable, with the modification of real data from global ocean observing stations. The data is suitable for the simulation of the wave regime in the study area. The results of the wave field were used to produce input data for the hydraulic model MIKE 21 HD.

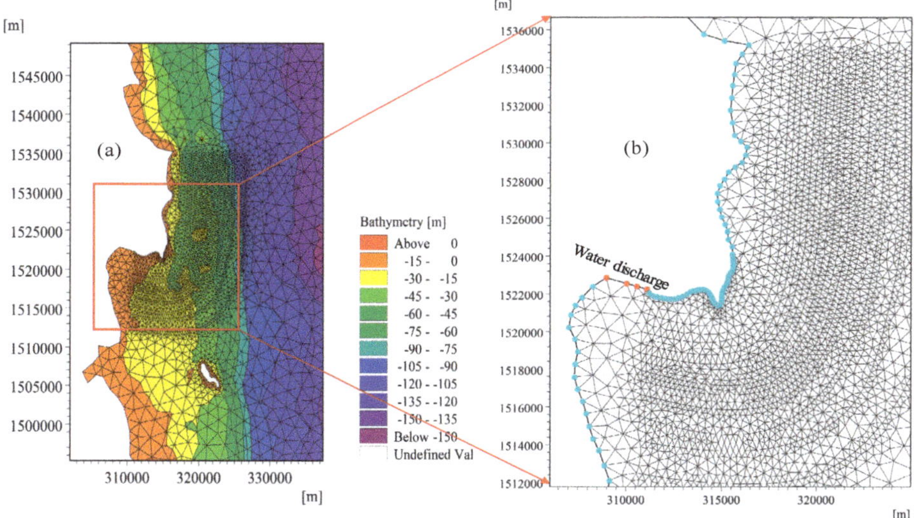



**Figure 3.** (**a**) 2D visualization bathymetry of wave mesh; (**b**) Detailed mesh of study area.

### 3.1.2. Calibration and Validation of Spectral Wind-Wave in 2D Model

In the seaside of Nhon Hai ward, the sea is relatively deep, the slope of the shoreline is steep, and the 20 m isobath is 15 km from the shore. In the study area, the wave regime is suitable for the wind regime, and the wave regime is divided into two seasons. In the winter, the direction of the wave is northeast (NE) with an average height between 0.8 and 0.9 m, and it may range from 1.1 to 1.2 m during the first three months. The biggest wave varies from 4.0–4.5 m. In the summer, the direction of the wave is south west (SW) and southeast (SE). The average height ranges from 0.6 to 0.7 m. The largest wave can be 4.0 m.

The calibration and validation of spectral wind-waves in a 2D model was undertaken using observed wave height data at Ly Son station during 1–30 September 2016 and 1–31 October 2016, respectively. The calculation of the wave characteristics showed that the simulated and observed wave heights were in good agreement in terms of the vibration amplitude, absolute value, and tidal phases during both the calibration and validation processes (Figure 4). The Nash–Sutcliffe efficient (NSE) [49] values ranged from 0.91 to 0.92 in the calibration and validation processes. Root mean square error (RMSE)-observations standard deviation ratio (RSR) values ranged from 0.27–0.3 in both calibration and validation. Percent bias (PBIAS) values for calibration and validation ranged from −5.6% to −3.8% (Percent Bias (PBIAS) < ± 10%) [50]. During the southwest monsoon, the 2D SW model simulated the general trend of the wave peaks well. The parameters obtained in calibration were maintained in the validation process with the bottom friction, based on Nikuradse roughness, ranged from 0.038–0.04 m. Dissipation coefficients white-capping (Cdis and Delta dis) were 4.5 and 0.5, respectively. The wave breaking coefficient ranged from 0.68–0.73. The validation results showed that the 2D SW model had successfully simulated the wave propagation in the study area (Figure 5). The parameters were reliable enough to conduct the wave field simulations under different scenarios in the study area.

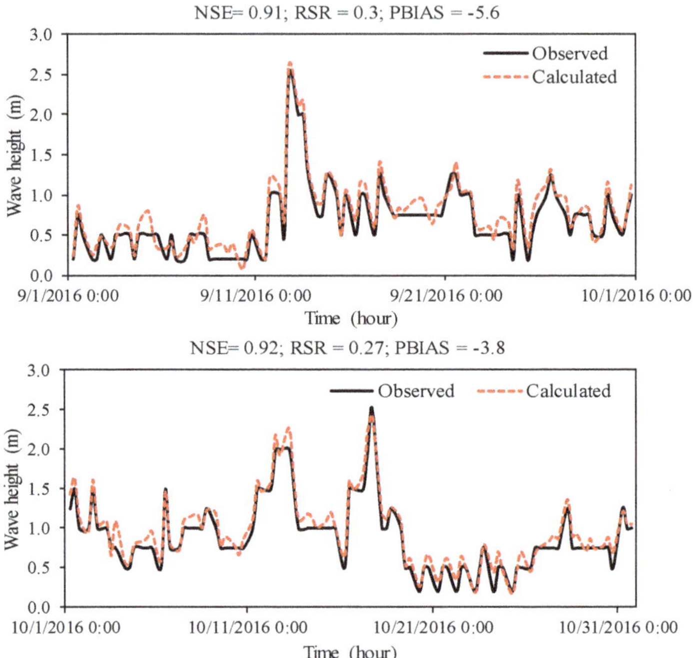

**Figure 4.** Calibration and validation of the wave height in 2D SW model at Ly Son station.

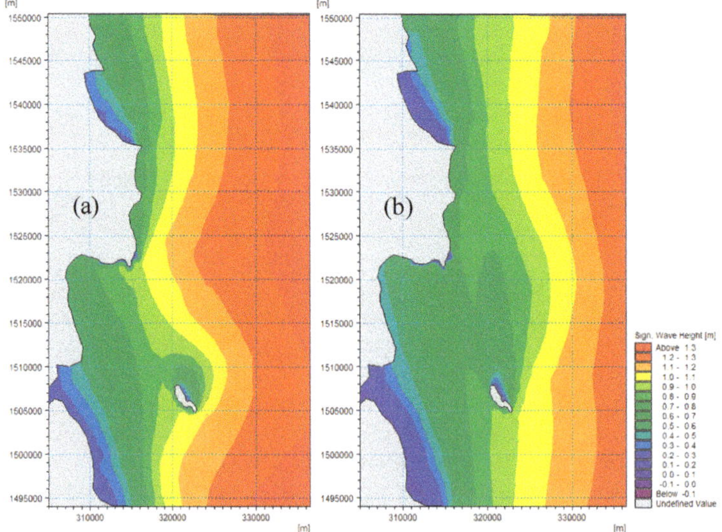

**Figure 5.** The results of wave field: (**a**) low tide and (**b**) high tide.

*3.2. Calibration and Validation of Hydrodynamic in 2D Model*

Calibration and validation of hydrodynamics in the 2D model used an observed water level tide at Quy Nhon station during 1–30 September 2016 and 1–31 October 2016, respectively. In this study, Nash–Sutcliffe efficient (NSE), Percent bias (PBIAS), and RMSE-observations standard deviation ratio (RSR) were used to evaluate the observed and simulated water levels [49,50]. The results of the calculated and observed water levels were in good agreement in terms of the vibration amplitude, absolute value, and tidal phases during both the calibration and validation processes (Figure 6). The NSE value for calibration and validation of the water level at Quy Nhon station ranged from 0.84–0.86. The maximum tide peak error in calibration and validation were 0.14 m and 0.16 m, respectively. According to the guidelines for model evaluation [50], the 2D model simulated the streamflow trends very well, as demonstrated by the statistical results, which were in agreement with the graphical results. The RSR values ranged from 0.32–0.44 for both calibration and validation. The PBIAS values ranged from −4.7 to −3.5 for both calibration and validation. The average magnitude of the simulated values had a good performance rating (PBIAS < ± 10%) for both calibration and validation. The simulation results for the streamflow regime using the 2D model were very good in terms of performance ratings, as revealed by NSE, RSR, and PBIAS. The calibration parameters used in the model validation process were the bed resistance Manning coefficients, which varied from 52–57 (m$^{1/3}$/s), the eddy viscosity coefficient (Smagorinsky) [51–53], which ranged from 0.28–0.32 (m$^2$/s), and wind friction, which ranged from 0.0018–0.002. The time step was 72 s. The last parameters were used to simulate mud transport in the dumping zone of the 2D model under different scenarios.

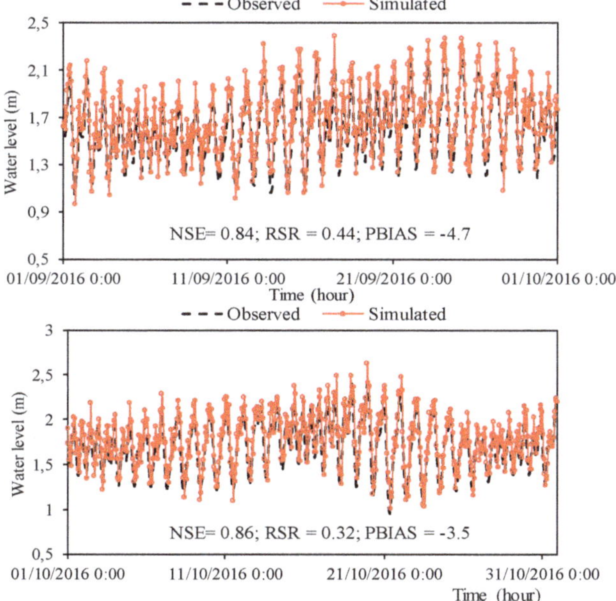



**Figure 6.** Calibration and validation of water level at Quy Nhon station.

The model results showed that the flow field in the study area was strongly impacted by topography, flow in the river, wind and wave fields (Figure 7). The flow field has tremendous fluctuations in both direction and speed. This means that the flow field can impact the sedimentation process differently in the various scenarios. To avoid this, the study decoupled the hydrodynamic module and mud transport module in the 2D model to simulate the sedimentation of the submerged zone.

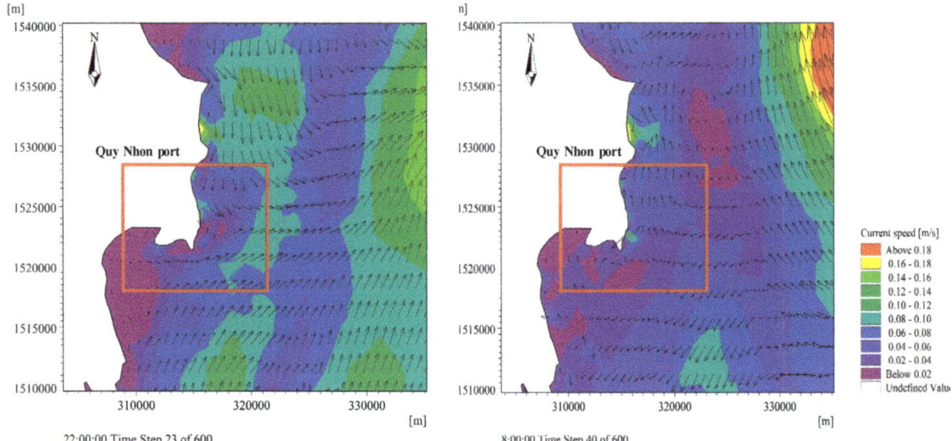


**Figure 7.** The flow field results at a closer view of Quy Nhon port.

### 3.3. Result of Mud Transport (MT) in 2D Model

The scenarios tested in the mud transport module are as below:

Scenario 1 (S1): Using the reanalysis of wind data from September and October 2017 produced from the NOAA GFS model for the study area.

Scenario 2 (S2): Calm wind

Scenario 3 (S3): Using the wind direction from September and October 2017, and additionally, the wind velocity reaches Beaufort level 4, with a wind speed of 6,7 m/s.

The simulation time was from 10 September 2017 to 30 September 2017, the time step for the simulation was 72 s, and the iteration was 60,000 steps.

Scenario 1 (S1): This scenario applied the reanalysis of wind data in September and October 2017 produced from the NOAA GFS model for the study area. The results of the hydraulic simulation demonstrated that streamflow velocity fluctuated from 0.08–0.72 m/s in the dredged area, streamflow velocity fluctuated from 0.08–0.16 m/s in the dredged area, the tidal flow in high tide could reach 0.36 m/s, and the flow velocity in the dredged area ranged from 0.12–0.28 m/s in the southeast (Figure 8a,b).

The hydraulic results affected the diffusion of total suspended solids (TSS) in the 2D MT model. Based on the results, the concentration of floating sludge under calm wind was 0.08 kg/m$^3$, which is below the acceptable limit (the tolerance limit of QCVN 10-MT:2015/BTNMT [54] is 0.5 kg/m$^3$). At low tide, the concentration was higher, at 0.0072 kg/m$^3$ diffused in an area of 6 km$^2$, but at high tide, the concentration was greater than 0.0045 kg/m$^3$ in an area of 8 km$^2$ (Figure 8c,d). However, this concentration was much lower than the acceptable limit; therefore, the dumping of dredged materials in this area was unlikely to increase the concentration of suspended material to beyond the acceptable limit.

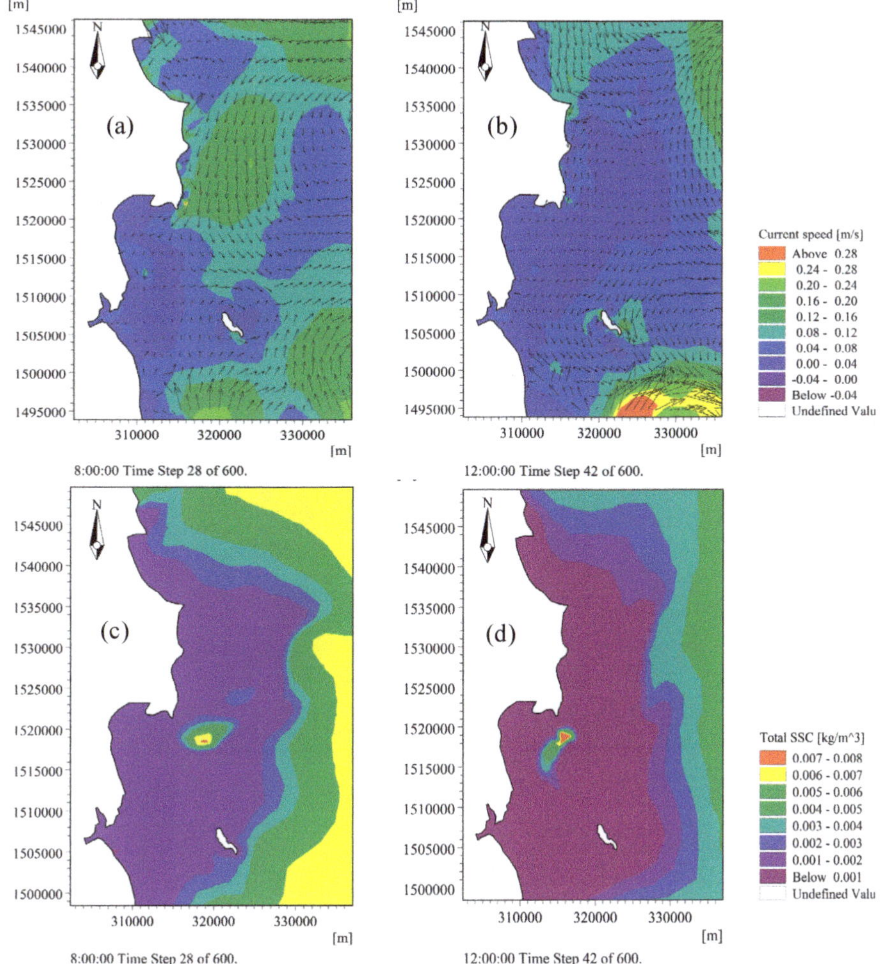

**Figure 8.** The simulation results of S1. Streamflow field during (**a**) low tide; (**b**) high tide. Distribution of suspended sediment concentration (SSC) during (**c**) low tide; (**d**) high tide.

Scenario 2 (S2): The simulation in calm wind showed that during peak and low tides, the flow velocity was smaller than 0.03 m/s. In the dredged area, flow velocity was in the southwest during the peak tide and in the west during low tide. In the ebb tide, flow velocity was northeast, at 0.04–0.36 m/s. At high tide, the flow was upstream, southwest, at 0.03–0.05 m/s (Figure 9a,b).

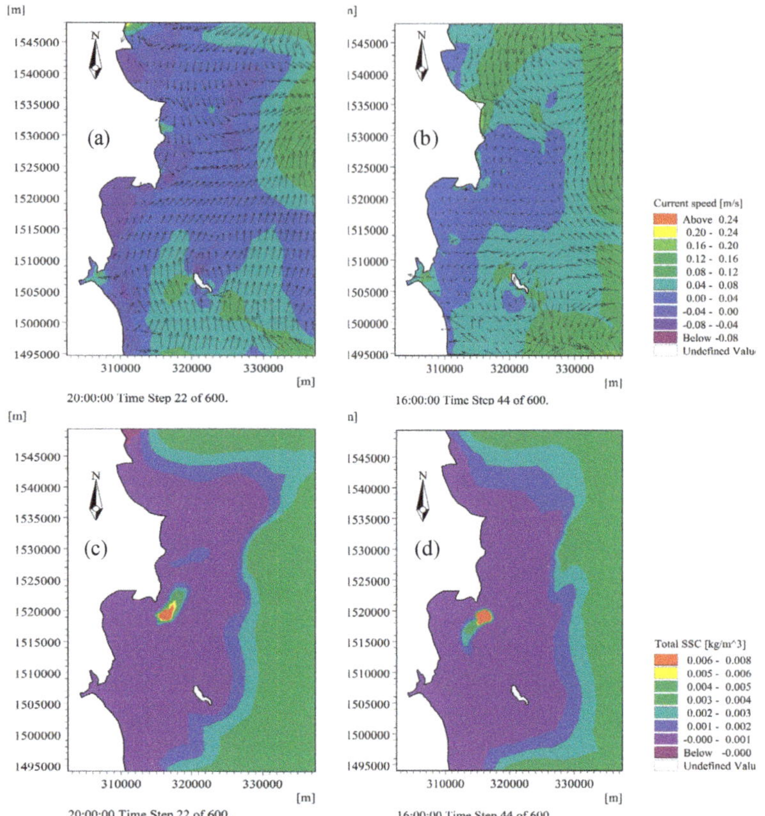

**Figure 9.** The simulation results of S2. Streamflow field during (**a**) low tide; (**b**) high tide. Distribution of suspended sediment concentration (SSC) during (**c**) low tide; (**d**) high tide.

The streamflow in the dredged area was quite stable, and the velocity was similar in both high and low tides. Therefore, the effect of streamflow on the concentration of total suspended solids was minimal. The suspended material moved to the northeast in the ebb tide and moved upstream to the southwest in the high tide. As a result, the distribution of the suspended material was an elliptical shape with the northeast–southwest axis. The simulation results showed that the maximum concentration of suspended sludge was under the acceptable limit [54] (the acceptance of QCVN 10-MT:2015/BTNMT is 0.5 kg/m$^3$). The concentration of suspended sludge was greater than 0.004 kg/m$^3$, spread over a 3 km$^2$ area (Figure 9c,d). However, this concentration was much smaller than the acceptable limit, so dumping dredged materials in this area is unlikely to adversely increase the concentration of suspended material.

Scenario 3 (S3): According to the simulated wind velocity during September and October 2017, the wind blew from the north to the south. At the dredged area, the direction of streamflow was southwest, and the current speed ranged from 0.08–0.24 m/s. At the outside of Hon Kho, the current speed could reach 0.6–0.75 m/s. During the ebb tide, the direction of run-off was similar, from the north to the south. At the dredged area, the streamflow was southwest, with speeds from 0.15–0.45 m/s. The streamflow outside Hong Kho extended to the south, with speed from 0.32–0.48 m/s (Figure 10a,b). The simulation of sediment sludge showed that the concentration was less than 0.05 kg/m$^3$ (the acceptable limit of QCVN 10-MT:2015/BTNMT [54]). At low tide, the concentration of suspended material was greater than 0.004 kg/m$^3$, spread over a 9 km$^2$ area, and during high tide, the concentration was 0.004 kg/m$^3$ over a 6 km$^2$ area (Figure 10c,d). However, this concentration was much lower than the acceptable limit, so the dump of dredged material in this area was unlikely to adversely increase the concentration of suspended material.

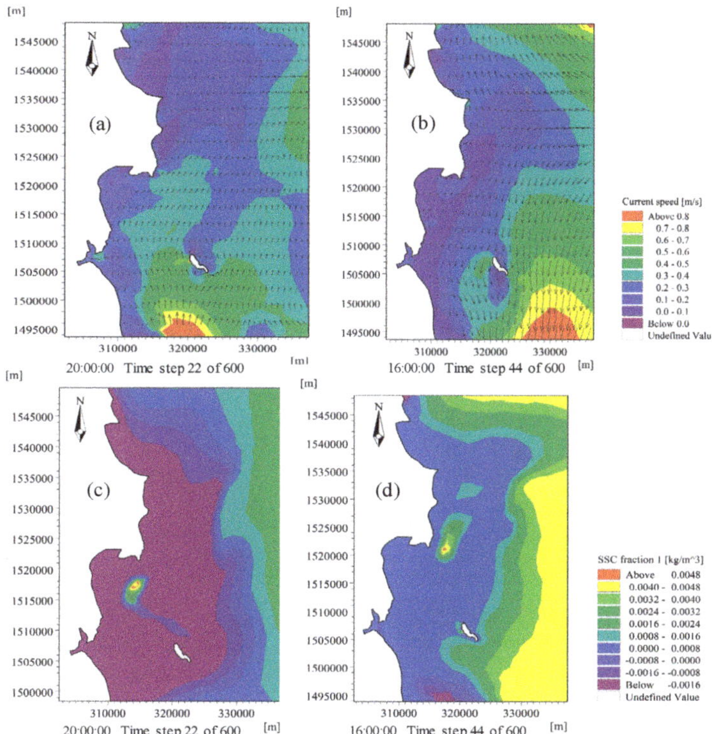

**Figure 10.** The simulation results of S3. Streamflow field during (**a**) low tide; (**b**) high tide. Distribution of suspended sediment concentration (SSC) during (**c**) low tide; (**d**) high tide.

### 3.4. Results of Particle Tracking (PT) in 2D Model

The impact of the dumped spoils on water quality of Nhon Hai commune, Quy Nhon resulted from the concentration of suspended material. The landfill of dredged material was 2.6 km to the south of the Nhon Hai commune (Figure 1). The dredged material, transported to the landfill by a hopper barge, was dumped in the following steps: The bottom door of the hopper barge was opened, and the material fell out of the deck and plunged into the sea. To evaluate the environmental impacts of dredge deposits, the study undertook a 2D model simulation, assuming that the dredged material was transported by a hopper barge at a rate of 2304 m$^3$/day; the concentration of suspended solids at the dumping position was 0.04 m$^3$/s and 0.50% of particles split from the dumped materials. MIKE 21 PT was decoupled from the hydraulic MIKE 21 HD. The simulated streamflow, as well as its momentum, would affect the concentration of the material. The results of the simulation of the three scenarios: Northeast (S1), calm wind (S2), and southwest (S3), are shown in Figure 11. The parameters of particle tracking (PT) used to simulate these scenarios in the 2D model are presented in Table 3. The distribution of sediment and the transmission distance of sediment particles are presented in Figure 11. After dumping, the seabed level in scenario 1 (S1) did not increase considerably (0.04 m); however, the seabed level in scenario 3 (S3) increased by 0.1 m to 0.2 m, and the seabed level increased by 0.05 m in the calm wind scenario (S2). The results of MIKE 21 PT show the following changes in seabed at the dumping position:

- In the case of calm wind (S2) and northeast wind (S1), the changes in seabed level due to the dumping of spoils were relatively similar.
- The increase in seabed level due to the dumping of spoils over a period of approximately 180 days was between 0.08–0.16 m over 0.0285 km$^2$.

As for the movement of particles at the dredged position, the element block moved to under the dredging position approximately 2.4 km in S1, and above the dredging position in S2 approximately 1.3 km. In S3, the element block moved quite far from the dredging position, approximately 2.9 km. Therefore, the results of the simulation were qualified, as the dredging position was situated far from the sea; therefore, it is significantly affected by the direction and velocity of wave and wind in the dredging position.

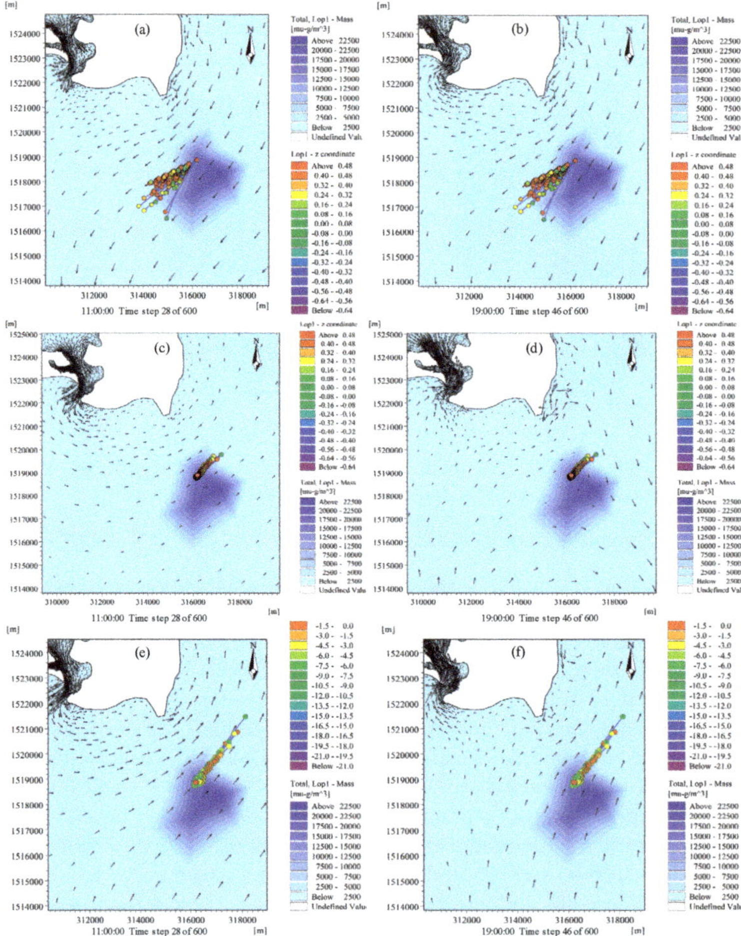

**Figure 11.** The simulation results of transferring sedimentation position in low tide and high tide in S1 (**a,b**); S2 (**c,d**); S3 (**e,f**).

**Table 3.** The parameters of particle tracking (PT) in 2D model.

| No. | Parameter | Value | Unit |
|---|---|---|---|
| 1 | Flux | 8.3 | kg/s |
| 2 | Number of particles per time step | 20 | Integer |
| 3 | Decay Rate | 0.05 | /sec |
| 4 | Settling velocity data | 0.1 | m/s |
| 5 | Settling parameters | | |
| 5.1 | Minimum concentration for | 0.01 | $kg/m^3$ |
| 5.2 | Maximum concentration for | 10 | $kg/m^3$ |
| 5.3 | Alpha | 1.01 | |
| 6 | Dispersion coefficient formulation | 0.05 | $m^2/s$ |

## 4. Conclusions and Discussion

The study simulates and evaluates the effects of dumping dredged spoils at Quy Nhon Port, which is one of the largest ports in Central Viet Nam. The dredging and maintenance of Quy Nhon Port play an important role in the transportation, evaluation and selection of dredging position and dredging volume. This study applies a 2D numerical model to simulate the propagation of contaminants in dredge materials to the marine environment, with the assumption that the total volume of dredged materials from Quy Nhon Port is transported by barges to the dumping location.

The reanalysis of wind data from the NOAA GFS model, with a high resolution of 0.5 degrees and a time span of 3 h, was used to simulate the wave-wind propagation in the 2D model. The simulation results were from 1 September 2017 to 31 October 2017. The calculated results of wave characteristics showed that the simulated and observed wave height were in good agreement in terms of the vibration amplitude, absolute value, and tidal phases in both the calibration and validation processes. The parameters were reliable enough to conduct wave field simulations under different scenarios in the study area. Calibration and validation of hydrodynamics in the 2D model used an observed water level tide at Quy Nhon station during 1–30 September 2016 and 1–31 October 2016, respectively. The results of the calculated and observed water levels were in good agreement in terms of the vibration amplitude, absolute value, and tidal phases, during both the calibration and validation processes.

The results of hydrodynamics and spectral wind-wave in the 2D model were used to simulate three scenarios using the mud transport module in the 2D model. The simulation results showed that the maximum concentration of suspended sludge was below the acceptable limit (the acceptance of [54] QCVN 10-MT:2015/BTNMT is 0.5 $kg/m^3$) for all three scenarios. The study used the particle tracking (PT) module in the 2D model to evaluate the environmental impact of dredge materials. The dredged materials were transported by hopper barge at 2304 $m^3$/day; the concentration was 0.04 $m^3$/s at the dredging position and 0.5% particles split from the dumped materials. The simulated results of the 2D PT model showed that the element block in scenario 3 moved the furthest from the dredging position, i.e., approximately 2.9 km. The results of the simulation were qualified, as the dredging position was situated far from the sea; therefore, it was significantly affected by the direction and velocity of wave and wind at the dredging position. For a future study, it is recommended that the uncertainty of the modeling needs is reduced. The following aspects may be improved:

- The computation time steps of the 2D model are currently large, although it takes approximately 4 to 5 h for one simulation of the 2D SW, HD, MT, PT model. A smaller time step can refine the results and provide more clarity, especially for sedimentation and particle tracking.
- The number of calibration and validation locations in the 2D models are few. It may not be enough to evaluate the adequacy of the model in the study area.

- The results of the 2D PT model only evaluate the distribution concentration and distance transmission of particles after the dredged materials. The results will be more useful and expanded if these are combined with the survey deposit patterns in the area surrounding the dump site.

**Author Contributions:** Conceptualization, D.Q.T. and J.K.; Methodology, D.Q.T.; Software, D.Q.T. and N.C.D.; Calibration and validation, D.Q.T. and N.C.D.; Formal analysis, J.K.; Investigation, N.C.D.; Data curation, D.Q.T.; Writing–original draft preparation, D.Q.T. and N.C.D.; Writing–review and editing, J.K.; Visualization, D.Q.T.

**Funding:** The authors gratefully acknowledge the financial support from the Ton Duc Thang University and the permission to use its facilities to perform the study.

**Conflicts of Interest:** The authors declare no conflict of interest.

## References


1. Bolam, S.G.; Rees, H.L. Minimizing impacts of maintenance dredged material disposal in the coastal environment: A habitat approach. *Environ. Manag.* **2003**, *32*, 171–188. [CrossRef]
2. Simonini, R.; Ansaloni, I.; Cavallini, F.; Graziosi, F.; Iotti, M.; N'Siala, G.M.; Mauri, M.; Montanari, G.; Preti, M.; Prevedelli, D. Effects of long-term dumping of dredged material on macrozoobenthos at four disposal sites along the Emilia-Romagna coast (Northern Adriatic Sea, Italy). *Mar. Pollut. Bull.* **2005**, *50*, 1595–1605. [CrossRef] [PubMed]
3. Leipe, T.; Kersten, M.; Heise, S.; Pohl, C.; Witt, G.; Liehr, G.; Zettler, M.; Tauber, F. Ecotoxicity assessment of natural attenuation effects at a historical dumping site in the western Baltic Sea. *Mar. Pollut. Bull.* **2005**, *50*, 446–459. [CrossRef] [PubMed]
4. Ware, S.; Bolam, S.G.; Rees, H.L. Impact and recovery associated with the deposition of capital dredgings at UK disposal sites: Lessions for future licensing and monitoring. *Mar. Pollut. Bull.* **2010**, *60*, 79–90. [CrossRef]
5. Kapsimalis, V.; Panagiotopoulos, I.; Kanellopoulos, T.; Hatzianestis, I.; Antoniou, P.; Anagnostou, C. A multi-criteria approach for the dumping of dredged material in the Thermaikos Gulf, Northern Greece. *J. Environ. Manag.* **2010**, *91*, 2455–2465. [CrossRef]
6. Bellas, J.; Nieto, Ó.; Beiras, R. Integrative assessment of coastal pollution: Development and evaluation of sediment quality criteria from chemical contamination and ecotoxicological data. *Cont. Self Res.* **2011**, *31*, 448–456. [CrossRef]
7. Aarninkhof, S.; Luijendijk, A. Safe disposal of dredged material in anenvironmentally sensitive environment. *Port Technol. Int.* **2010**, *47*, 39–45.
8. Erftemeijer, P.L.A.; Riegl, B.; Hoeksema, B.W.; Todd, P.A. Environmental impacts of dredging and other sediment disturbances on corals. *Mar. Pollut. Bull.* **2012**, *64*, 1737–1765. [CrossRef]
9. Maren, D.S.; Kessel, T.; Cronin, K.; Sittoni, L. The impact of channel deepening and dredging on estuarine sediment concentration. *Cont. Self Res.* **2015**, *95*, 1–14. [CrossRef]
10. Staniszewska, M.; Boniecka, H. Dangerous compounds in the dredged material from the sea—Assessment of the current approach to the evaluation of contaminations based on the data from the Polish coastal zone (the Baltic Sea). *Mar. Pollut. Bull.* **2018**, *130*, 324–334. [CrossRef]
11. Frenzel, P.; Bormann, C.; Lauenburg, B.; Bohling, B.; Bartholdy, J. Environment impact assessment of sediment dumping in the southern Baltic Sea using meiofaunal indicators. *J. Mar. Syst.* **2009**, *75*, 430–440. [CrossRef]
12. Boniecka, H.; Staniszewska, M.; Sapota, G.; Dembska, G.; Suzdalev, S. Przewodnik do wyznaczania nowych miejsc pod klapowiska. In *Guideline for the Location of New Dumping Sites, ECODUMP*; Maritime Institute of Gdańsk: Gdańsk, Poland, 2014; p. 36.
13. Staniszewska, M.; Boniecka, H.; Gajecka, A. Prace pogłębiarskie w polskiej strefie przybrzeżnej-aktualne problemy (Dredging works in the polish coastal zone-actual problems). *Inż. Ekol.* **2014**, *40*, 157–172. [CrossRef]
14. Staniszewska, M.; Boniecka, H.; Cylkowska, H. Dredging works in the Polish open sea ports as an anthropogenic factor of development of sea coastal zones. *Bull. Marit. Inst. Gdan.* **2016**, *31*, 180–187. [CrossRef]
15. Staniszewska, M.; Boniecka, H. The Environmental Protection Aspects of Handling Dredged Material. *BMI Bull. Marit. Inst. Gdan.* **2015**, *30*, 51–58. [CrossRef]
16. Staniszewska, M.; Boniecka, H. Managing dredged material in the coastal zone of the Baltic Sea. *Environ. Monit. Assess.* **2017**, *189*, 46. [CrossRef]

17. Arizaga, J.; Amat, J.A.; Ganuzas, M.M. The negative effect of dredging and dumping on shorebirds at a coastal wetland in northern Spain. *J. Nat. Conserv.* **2017**, *37*, 1–7. [CrossRef]

18. Chen, C.F.; Chen, C.W.; Ju, Y.R.; Kao, C.M.; Dong, C.D. Impact of disposal of dredged material on sediment quality in the Kaohsiung Ocean Dredged Material Disposal Site, Taiwan. *Chemosphere* **2018**, *191*, 555–565. [CrossRef] [PubMed]

19. Erftemeijer, P.L.A.; Lewis, R.R., III. Environmental impacts of dredging on seagrasses: A review. *Mar. Pollut. Bull.* **2006**, *52*, 1553–1572. [CrossRef] [PubMed]

20. Hoeksema, B.W. Delineation of the Indo-Malayan Centre of maximum marine biodiversity: The Coral Triangle. In *Biogeography, Time and Place. Distributions. Barriers and Islands*; Renema, W., Ed.; Springer: Dordrecht, The Netherlands, 2007; pp. 117–178.

21. Jones, R.; Browne, P.B.; Fisher, R.; Klonowski, W.; Slivkoff, M. Assessing the impacts of sediments from dredging on corals. *Mar. Pollut. Bull.* **2016**, *102*, 9–29. [CrossRef]

22. Staniszewska, M.; Boniecka, H. Badania przesiewowe w ocenie stopnia zanieczyszczenia urobku. Screening methods for dredged material contamination assessment. *BMI Bull. Marit. Inst. Gdan.* **2016**, *31*, 80–87. [CrossRef]

23. Foster, G.; Annan, J.D.; Jones, P.D.; Mann, M.E.; Mullan, B.; Renwick, J.; Salinger, J.; Schmidt, G.A.; Trenberth, K.E. Comment on "Influence of the Southern Oscillation on tropospheric temperature" by J. D. McLean, C. R. de Freitas, and R. M. Carter. *J. Geophys. Res.* **2010**, *115*, D09110. [CrossRef]

24. Quy Nhon Port. 2012. Available online: http://www.quinhonport.com.vn/en/gioi-thieu/17/gioi-thieu-chung- (accessed on May 2011).

25. Hervouet, J.M. *Hydrodynamics of Free Surface Flows: Modelling with the Finite Element Method*; John Wiley & Sons Ltd.: Chichester, UK, 2007.

26. Maerker, C.; Malcherek, A. *The Simulation Tool DredgeSim—Predicting Dredging Needs in 2- and 3-Dimensional Models to Evaluate Dredging Strategies*; Dittrich, A., Koll, K., Aberle, J., Geisenhainer, P., Eds.; Available online: https://hdl.handle.net/20.500.11970/99824 (accessed on 26 November 2014).

27. Guerrero, M.; di Federico, V.; Lamberti, A. Calibration of a 2-D morphodynamic model using water-sediment flux maps derived from an ADCP recording. *J. Hydroinform.* **2013**, *15*, 813–828. [CrossRef]

28. Paarlberg, A.J.; Guerrero, M.; Huthoff, F.; Re, M. Optimizing Dredge-and-Dump Activities for River Navigability Using a Hydro-Morphodynamic Model. *Water* **2015**, *7*, 3943–3962. [CrossRef]

29. Shukla, V.K.; Konkane, V.D.; Nagendra, T.; Agrawal, J.D. Dredged Material Dumping Site Selection Using Mathematical Models. *Procedia Eng.* **2015**, *116*, 809–817. [CrossRef]

30. Ramli, A.Y.; de Lange, W.; Bryan, K.; Mullarney, J. Coupled Flow-Wave Numerical Model in Assessing the Impact of Dredging on the Morphology of Matakana Banks. In Proceedings of the Australasian Coasts & Ports Conference 2015, Auckland, New Zealand, 15–18 September 2015.

31. Morais, M.; Dibajnia, M.; Lu, Q.; Fournier, C. 3D modelling of combined dredge and disposal plumes dispersion, ponta da madeira, Brazil. In Proceedings of the WEDA XXXIV Technical Conference & TAMU 45 Dredging Seminar, Toronto, ON, Canada, June 2014.

32. Gaeta, M.G.; Bonaldo, D.; Samaras, A.G.; Carniel, S.; Archetti, R. Coupled Wave-2D Hydrodynamics Modeling at the Reno River Mouth (Italy) under Climate Change Scenarios. *Water* **2018**, *10*, 1380. [CrossRef]

33. Davies, S.; Mirfenderesk, H.; Tomlinson, R.; Szylkarski, S. Hydrodynamic, water quality and sediment transport modeling of estuarine and coastal waters on the gold coast Australia. *Proc. J. Coast. Res.* **2009**, *56*, 937–941.

34. Doan, Q.T.; Nguyen, C.D.; Chen, Y.C. Trajectory Modelling of Marine Oil Spills: Case Study of Lach Huyen Port, Vietnam. *Lowl. Technol. Int.* **2013**, *15*, 41–51. [CrossRef]

35. Menendez, A.N.; Badano, N.D.; Lopolito, M.F.; Re, M. Water quality assessment for a coastal zone through numerical modeling. *J. Appl. Water Eng. Res.* **2013**, *1*, 8–16. [CrossRef]

36. Shi, Z. Estuaries: Dynamics, mixing, sedimentation and morphology. *J. Coast. Res.* **2010**, *26*, 586–587. [CrossRef]

37. Fossati, M.; Piedra-Cueva, I. A 3D hydrodynamic numerical model of the Río de la Plata and Montevideo's coastal zone. *Appl. Math. Model.* **2013**, *37*, 1310–1332. [CrossRef]

38. Doan, Q.T.; Nguyen, C.D.; Chen, Y.C.; Pawan, K.M. Application of Environmental Sensitivity Index (ESI) Maps of Shoreline for the Coastal Oil Spills: Case Study of Cat Ba Island, Vietnam. *Environ. Earth Sci.* **2015**, *74*, 3433–3451. [CrossRef]

39. Suntoyo; Ikhwani, H.; Zikra, M.; Sukmasari, N.A.; Angraeni, G.; Tanaka, H.; Umeda, M.; Kure, S. Modelling of the COD, TSS, Phosphate and Nitrate Distribution Due to the Sidoardjo Mud Flow into Porong River Estuary. *Procedia Earth Planet. Sci.* **2015**, *14*, 144–151. [CrossRef]

40. Doan, Q.T.; Nguyen, T.M.L.; Nguyen, C.D. Using numerical modelling in the simulation of mass fish death phenomenon along the Central Coast of Vietnam. *Mar. Pollut. Bull.* **2018**, *129*, 740–749. [CrossRef]

41. Doan, Q.T.; Nguyen, T.M.L.; Tran, H.T.; Kandasamy, J. Application of 1D-2D Coupled Modeling in Water Quality Assessment: A Case Study in Ca Mau Peninsula, Vietnam. *Phys. Chem. Earth* **2018**. [CrossRef]

42. Komen, G.J.; Cavaleri, L.; Doneland, M.; Hasselmann, K.; Hasselmann, S.; Janssen, P.A.E.M. *Dynamics and Modeling of Ocean Waves*; Cambridge University Press: Cambridge, UK, 1994; p. 560.

43. Young, I.R. Wind-generated Ocean waves. In *Elsevier Ocean Engineering Book Series*; Bhattacharyya, R., McCormick, M.E., Eds.; 1999; Volume 2, Available online: https://www.elsevier.com/books/wind-generated-ocean-waves/young/978-0-08-043317-2 (accessed on 23 March 1999).

44. Martin, J.L.; McCutcheon, S.C. *Hydrodynamics and Transport for Water Quality Modeling*; Taylor & Francis Group: Boca Raton, London, 1998; 816p, Available online: https://www.crcpress.com/Hydrodynamicsand-Transport-for-Water-Quality-Modeling/Martin-McCutcheon/p/book/9780873716123 (accessed on 4 May 2018).

45. Krone, R.B. *Flume Studies of the Transport of Sediment in Estuarial Processes*; Final Report; Hydraulic Engineering Laboratory and Sanitary Engineering Research Laboratory, University of California: Barkeley, CA, USA, 1962.

46. Mehta, A.J.; Hayter, E.J.; Parker, W.R.; Krone, R.B.; Teeter, A.M. Cohesive sediment transport. *J. Hydraul. Eng.* **1989**, *115*, 1076–1093. [CrossRef]

47. Sikder, S.; Hossain, F. Assessment of the weather research and forecasting model generalized parameterization schemes for advancement of precipitation forecasting in monsoon-driven river basins. *J. Adv. Model. Earth Syst.* **2016**, *8*, 1210–1228. [CrossRef]

48. Doan, Q.T.; Nguyen, T.M.L.; Quach, T.T.T.; Tran, A.P.; Nguyen, C.D. Assessment of Water Quality in Coastal Estuaries Under Impact of Industrial Zone in Hai Phong, Vietnam. *Phys. Chem. Earth* **2019**, accepted.

49. Nash, J.E.; Sutcliffe, J.V. River flow forecasting through conceptual models parts I—A discussion of principles. *J. Hydrol.* **1970**, *10*, 282–290. [CrossRef]

50. Moriasi, D.N.; Arnold, J.G.; Van Liew, M.W.; Bingner, R.L.; Harmel, R.D.; Veith, T.L. Model evaluation guidelines for systematic quantification of accuracy in watershed simulations. *Trans. ASABE* **2007**, *50*, 885–900. [CrossRef]

51. Akira, Y. Eddy-viscosity-type subgrid-scale model with a variable Smagorinsky coefficient and its relationship with the one-equation model in large eddy simulation. *Phys. Fluids* **1991**, *3*, 2007–2009. [CrossRef]

52. Cui, G.X.; Xu, C.X.; Fang, L.F.; Shao, L.; Zhang, Z.S. A new subgrid eddy-viscosity model for large-eddy simulation of anisotropic turbulence. *J. Fluid Mech.* **2007**, *582*, 377–397. [CrossRef]

53. Romit, M.; Omer, S. Dynamic modeling of the horizontal eddy viscosity coefficient for quasigeostrophic ocean circulation problems. *J. Ocean Eng. Sci.* **2016**, *1*, 300–324. [CrossRef]

54. Circulars 67, QCVN 10-MT/BTNMT, National Technical Regulation on Marine Water Quality. 2015. Available online: https://thuvienphapluat.vn/van-ban/Tai-nguyen-Moi-truong/Thong-tu-67-2015-TT-BTNMT-quy-chuan-ky-thuat-quoc-gia-moi-truong-301666.aspx (accessed on 21 December 2015).

*Article*

# A Novel Remote Sensing Index for Extracting Impervious Surface Distribution from Landsat 8 OLI Imagery

**Hong Fang [1,2,3], Yuchun Wei [1,2,3,*] and Qiuping Dai [1,2,3]**

[1]  School of Geography, Nanjing Normal University, Nanjing 210023, China
[2]  Key Laboratory of Virtual Geographic Environment (Nanjing Normal University), Ministry of Education, Nanjing 210023, China
[3]  Jiangsu Center for Collaborative Innovation in Geographical Information Resource Development and Application, Nanjing 210023, China
*  Correspondence: weiyuchun@njnu.edu.cn; Tel.: +86-18651602916

Received: 25 April 2019; Accepted: 27 June 2019; Published: 28 June 2019

**Abstract:** The area of urban impervious surfaces is one of the most important indicators for determining the level of urbanisation and the quality of the environment and is rapidly increasing with the acceleration of urbanisation in developing countries. This paper proposes a novel remote sensing index based on the coastal band and normalised difference vegetation index for extracting impervious surface distribution from Landsat 8 multispectral remote sensing imagery. The index was validated using three images covering urban areas of China and was compared with five other typical index methods for the extraction of impervious surface distribution, namely, the normalised difference built-up index, index-based built-up index, normalised difference impervious surface index, normalised difference impervious index, and combinational built-up index. The results showed that the novel index provided higher accuracy and effectively distinguished impervious surfaces from bare soil, and the average values of the recall, precision, and F1 score for the three images were 95%, 91%, and 93%, respectively. The novel index provides better applicability in the extraction of urban impervious surface distribution from Landsat 8 multispectral remote sensing imagery.

**Keywords:** ratio-based impervious surface index; remote sensing information extraction; urban impervious surface; urbanisation

---

## 1. Introduction

Impervious surfaces are artificial surfaces that water cannot infiltrate to reach the soil, such as parking lots, streets, and highways [1]. Changes in land use and land cover caused by the expansion of impervious surfaces are likely to influence the regional climate [2]. Jacobson et al. and Huszar et al. pointed out that the impervious surface area and land surface temperature are positively correlated, and the expansion of impervious surfaces will result in the increase of the land surface temperature which may cause uneven distribution of urban heat and trigger the urban heat island effect [3,4].

Accurate and fast extraction methods for impervious surface distribution are important which is helpful for detecting regional environmental changes in urban areas and achieving sustainable urban development. They can be divided into two categories, namely, field surveys and remote sensing [1]. Field surveys can provide more detailed information on impervious surface distribution but are time-consuming, laborious, and difficult to apply to the assessment of large areas. In contrast, widely-used remote sensing methods can extract impervious surface distribution quickly, and sources of commonly used remote sensing images include the QuickBird [5], WorldView-2 [6], and Landsat satellites [7], etc. When extracting impervious surface distribution in a large urban area, Landsat imagery is often the primary choice because of its medium spatial resolution and free availability, while remote sensing imagery with fine spatial resolution is subject to economic constraints.

Remote sensing methods used in the extraction of impervious surface distribution include index methods, spectral mixture analysis (SMA), and classification methods. Firstly, index methods construct a feature index based on the spectral difference between impervious surfaces and other ground objects and have been a popular research focus because of their simplicity in calculation. Zha et al. constructed the normalised difference built-up index (NDBI) on the basis of the near-infrared and shortwave-infrared bands [7]. Xu proposed the index-based built-up index (IBI), which is based on the NDBI, soil-adjusted vegetation index (SAVI) and modified normalised difference water index (MNDWI) [8]. Xu constructed the normalised difference impervious surface index (NDISI) on the basis of the thermal band, MNDWI, near-infrared band, and shortwave-infrared band [9]. Wang et al. devised the normalised difference impervious index (NDII), which is based on the visible and thermal bands [10]. Sun et al. developed the combinational built-up index (CBI) by combining the first principal component (PC1), normalised difference water index (NDWI), and SAVI [11]. Secondly, SMA calculates the fractions of land cover types in a pixel and models a mixed spectrum as a combination of the spectra for pure types of land cover, and it includes linear SMA (LSMA) and nonlinear SMA (NLSMA). Voorde et al. utilised LSMA to extract the impervious surface distribution [12]. Zhao et al. revised LSMA using principal component analysis and the NDBI [13]. Roberts et al. proposed multiple-endmember spectral mixture analysis [14], which was improved by Fan et al. to achieve higher efficiency and precision [15]. Finally, classification methods divide pixels into impervious and permeable surfaces according to the features of the image, and commonly used classifiers include support vector machines (SVMs) [16], artificial neural networks (ANNs) [17], and random forest (RF) [18], etc. Tehrany et al. extracted impervious surface distribution in Malaysia by an SVM [19]. Patel and Mukherjee used the ANN method [20]. Pelletier et al. utilised the RF method to extract impervious surface distribution in France and compared it with the SVM method [21]. In general, SMA and classification methods are complex and time-consuming, and their accuracy is easily influenced by the quality of the selected samples, whereas index methods involve no additional parameters and are more easily applied [11].

Extraction methods of impervious surface distribution from Landsat imagery (Thematic Mapper (TM) and Landsat 7 Enhanced Thematic Mapper Plus (ETM+)) have often been validated using imagery obtained in summer, when the vegetation is lush and often distinguished from impervious surfaces easily [7–10]. However, remote sensing images are not always available in every area in summer, such as the east of China where summer is often rainy. Besides, the methods themselves have poor abilities to mask bare soil [10,11]. The quality of Landsat 8 Operational Land Imager (OLI) imagery is better than that of Landsat 5 TM and Landsat 7 ETM+ imagery, but the existing index methods for the extraction of impervious surface distribution may lose applicability because of the differences in the sensor, wavelength range, and image quantisation bit [22].

In this study, a ratio-based impervious surface index (RISI) was proposed to simplify the extraction process and mask permeable surfaces such as bare soil. The index was validated by three typical images covering urban areas of China in winter and summer and the results were compared with those of five other typical methods for the extraction of impervious surface distribution, namely, the NDBI, IBI, NDISI, NDII, and CBI.

## 2. Data and Methods

### 2.1. Data

In this study, three case images without cloud or snow downloaded in Geospatial Data Cloud [23] were used of which two were obtained in summer when vegetation is lush and one was obtained in winter when vegetation is withered and bare soil is widely distributed. The image sizes are 480 rows × 480 columns with a pixel size of 30 m, and their (5, 4, 3) false colour composites are shown in Figure 1.

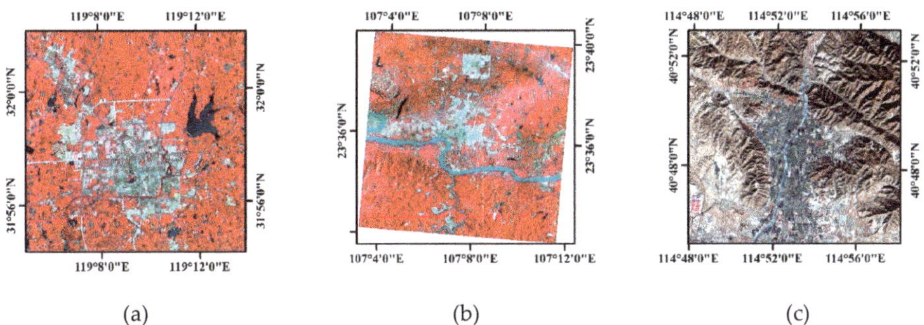

**Figure 1.** (5, 4, 3) false colour composite images with a linear stretch of 2%: (**a**) Case 1, 2017/7/21; (**b**) Case 2, 2014/7/23; (**c**) Case 3, 2013/12/13.

Typical ground objects in these case images include vegetation, bare soil, water body, and impervious surfaces, and present more representative and complexely. In the (5, 4, 3) false colour composite image with a 2% linear stretch, the vegetation is red in summer such as Cases 1 and 2 and easily distinguished from impervious surfaces because of their different spectral features; bare soil is exposed in winter, such as in Case 3, and is easily confused with impervious surfaces because of their similar spectral features.

*2.2. Extraction of Impervious Surface Distribution*

2.2.1. Proposed Method

According to the vegetation–impervious surface–soil (VIS) model [24], land cover consists mainly of vegetation (V), impervious surfaces (I) and bare soil (S), with the exception of water bodies. In remote sensing imagery, the largest difference in land cover is that between water bodies and land, and land can further be divided into impervious surfaces and permeable surfaces (vegetation and bare soil). Hence, the workflow can be divided into two steps, namely, masking water bodies and extracting impervious surfaces from land.

An impervious surface is an area without vegetation coverage and thus has a lower NDVI value [25]. Studies by Wang et al. and Mu et al. showed that impervious surfaces usually have higher values in the blue band [10,26]. Besides, Chen pointed out that the coastal band in Landsat 8 imagery can enhance information on impervious surfaces more significantly than the blue band, and the addition of the coastal band in the supervised classification of land use is beneficial for further distinguishing impervious surfaces from other ground objects [27].

In this study, a RISI was proposed for extracting impervious surface distribution on the basis of the coastal band and NDVI of a digital number (DN) image and is defined by Equation (1):

$$RISI = \frac{B1}{NDVI1} \tag{1}$$

where B1 and NDVI1 denote the coastal band and NDVI, respectively, after a 0–1 transformation.

The workflow comprises the following four steps:

(a) Water bodies are first masked;

(b) The coastal band undergoes a 0–1 transformation to obtain B1;

(c) The NDVI is calculated, and then a 0–1 transformation is performed to obtain NDVI1; and

(d) The RISI is calculated and then segmented using a threshold determined by the Otsu method [28].

1. Water Body Masking

The water body index (WI) is calculated using Equation (2) [29]:

$$WI = \frac{Green}{SWIR1} \tag{2}$$

where Green denotes the green band and SWIR1 denotes the shortwave-infrared band.

The value corresponding to the right valley of the WI histogram is manually fine-tuned as the segmentation threshold to mask water bodies.

2. 0–1 Transformation of the Coastal Band

The coastal band undergoes a 0–1 transformation to obtain B1 according to Equation (3):

$$T1 = \frac{T0 - min}{max - min} \tag{3}$$

where T0 and T1 denote the band before and after the 0–1 transformation according to Equation (3), respectively, and max and min are the maximum and minimum values of the band, respectively.

3. Calculation of the NDVI

The NDVI is calculated using Equation (4) [30]:

$$NDVI = \frac{NIR - Red}{NIR + Red} \tag{4}$$

where Red denotes the red band and NIR denotes the near-infrared band. Then, the NDVI undergoes a 0–1 transformation to obtain NDVI1 in an analogous way to Equation (3).

4. Calculation of the RISI

The RISI is calculated using Equation (1), and then its threshold is determined by the Otsu method and used to segment the image to obtain the impervious surface region.

The Otsu method is an adaptive threshold determination method and is widely used in image segmentation, which maximizes the class variance between objects and background [28].

2.2.2. Comparison of Methods

The NDBI, IBI, NDISI, NDII, and CBI methods were used for comparison. The segmentation thresholds of the NDBI and NDISI were both set to 0, which is the default threshold according to the authors who proposed these indices, whereas the thresholds of the other three indices were determined by the Otsu method here because the authors who proposed these indices determined the thresholds manually rather than setting a fixed threshold. In addition, in all the methods, water bodies were masked first so as to ensure the comparisons coincided.

(1) NDBI: Zha et al. proposed the NDBI for extracting impervious surface distribution [7]:

$$NDBI = \frac{SWIR1 - NIR}{SWIR1 + NIR} \tag{5}$$

(2) IBI: Xu constructed the IBI to extract impervious surface distribution [8]:

$$BI = \frac{NDBI - \frac{(SAVI + MNDWI)}{2}}{NDBI + \frac{(SAVI + MNDWI)}{2}} \tag{6}$$

where SAVI [31] is calculated as follows:

$$SAVI = \frac{(NIR - Red)(1 + L)}{NIR + Red + L} \tag{7}$$

where L is a correction factor that ranges from 0 to 1. In this study, L was set to 0.5, which can minimise noise due to soil [11,31].

(3) NDISI: Xu proposed the NDISI for extracting impervious surface distribution [9]:

$$NDISI = \frac{TIR - (MNDWI + NIR + SWIR1)/3}{TIR + (MNDWI + NIR + SWIR1)/3} \tag{8}$$

where TIR denotes the thermal band, which we replaced with the TIR1 band (10.6–11.19 μm) in Landsat 8 imagery. MNDWI is calculated as follows [32]:

$$MNDWI = \frac{Green - SWIR1}{Green + SWIR1} \tag{9}$$

(4) NDII: Wang et al. devised the NDII to extract impervious surface distribution [10]:

$$NDII = \frac{VIS - TIR}{VIS + TIR} \tag{10}$$

where VIS represents one of the red, green and blue bands. Wang pointed out that the accuracy of extraction is highest when VIS represents the red band [10], and hence we also used the red band to calculate the NDII.

(5) CBI: Sun et al. developed the CBI to extract impervious surface distribution [11]:

$$CBI = \frac{\frac{(PC1+NDWI)}{2} - SAVI}{\frac{(PC1+NDWI)}{2} + SAVI} \tag{11}$$

where NDWI is calculated as follows [33]:

$$NDWI = \frac{Green - NIR}{Green + NIR} \tag{12}$$

2.2.3. Method for Accuracy Assessment

Accuracy is determined in terms of precision, recall, and the F1 score.

According to their actual and predicted values, pixels are divided into true positives (TP), false positives (FP), true negatives (TN), and false negatives (FN), and then the precision (P), recall (R), and F1 score are calculated using Equations (13)–(15) [34]:

$$P = \frac{TP\_N}{TP\_N + FP\_N} \times 100\% \tag{13}$$

where TP_N and FP_N are the total numbers of pixels in TP and FP, respectively.

$$R = \frac{TP\_N}{TP\_N + FN\_N} \times 100\% \tag{14}$$

where FN_N is the total number of pixels in FN.

$$F1 = \frac{2 \times P \times R}{P + R} \times 100\% \tag{15}$$

The closer the F1 score is to 1, the higher the accuracy of the extraction of impervious surface distribution is.

## 3. Results

### 3.1. Spectral Features of Impervious Surfaces, Bare Soil, and Vegetation

Typical samples of impervious surfaces, bare soil, and vegetation were selected from each case image after water bodies were masked, and the mean spectral values for each type of sample were calculated to obtain their spectral curves (Figure 2). These show that impervious surfaces had similar spectral features to those of bare soil. The DN value of impervious surfaces in the coastal band was higher than those of vegetation and bare soil, and vegetation had the largest difference between the DN values in the red and near-infrared bands, followed by bare soil. In the TIR1 band, the DN value of impervious surfaces was the highest in Case 2, whereas in Cases 1 and 3 the DN value of bare soil was the highest.

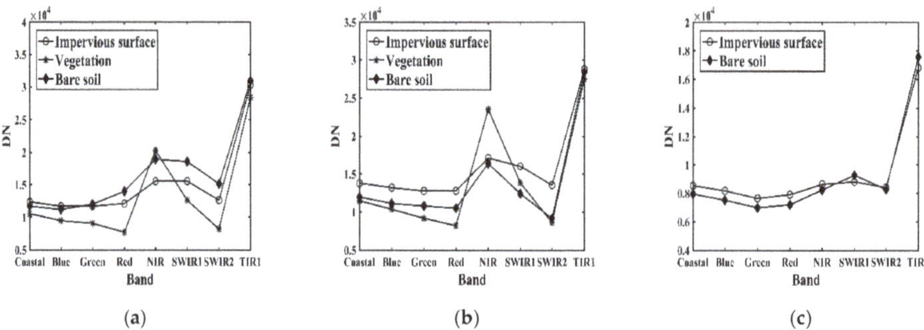

Figure 2. Spectral features of impervious surfaces, bare soil and vegetation in images after water bodies were masked: (**a**) Case 1; (**b**) Case 2; (**c**) Case 3.

### 3.2. Results of Extraction of Impervious Surface Distribution

The ground truth for impervious surface distribution was determined by visual interpretation in each case image and is shown in Figure 3:

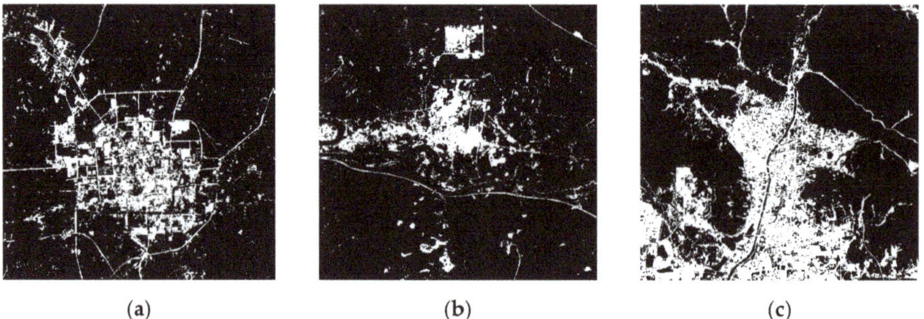

Figure 3. Ground truth for impervious surface distribution in the representative images: (**a**) Case 1; (**b**) Case 2; (**c**) Case 3.

Impervious surface distribution extracted by the proposed method is shown in Table 1. We can see that water bodies were masked well by the WI. The RISI effectively distinguished impervious surfaces from bare soil and vegetation, especially the large amount of bare soil in Case 3.

**Table 1.** The impervious surface distribution extracted by the proposed index.

| Case | (5, 4, 3) False Colour Composite Images with a Linear Stretch of 2% | Images after Water Bodies were Masked by the WI | The Impervious Surface Distribution Extracted by the RISI |
|---|---|---|---|
| 1 | | | |
| 2 | | | |
| 3 | | | |

The impervious surface distribution extracted by six methods are shown in Figures 4–6. The segmentation thresholds of the NDBI and NDISI were both set to 0, and the thresholds of the other methods were determined by the Otsu method. In general, the impervious surfaces extracted by the five reference methods were still mixed with other ground objects even though water bodies were masked. In Case 3, bare soil were seriously mis-extracted.

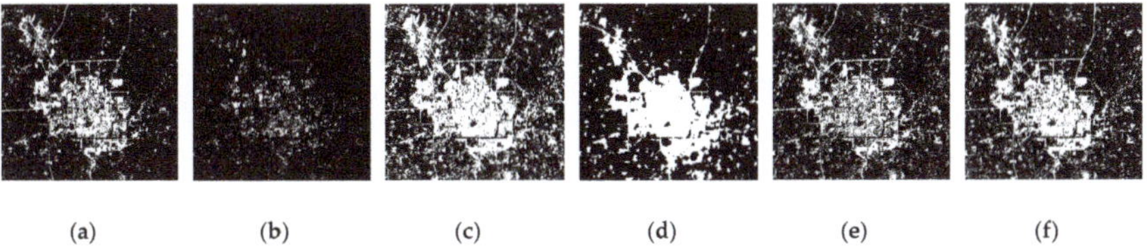

(a)    (b)    (c)    (d)    (e)    (f)

**Figure 4.** The impervious surface distribution in Case 1 extracted by six methods after water bodies were masked: (**a**) RISI; (**b**) NDBI; (**c**) IBI; (**d**) NDISI; (**e**) NDII; (**f**) CBI.

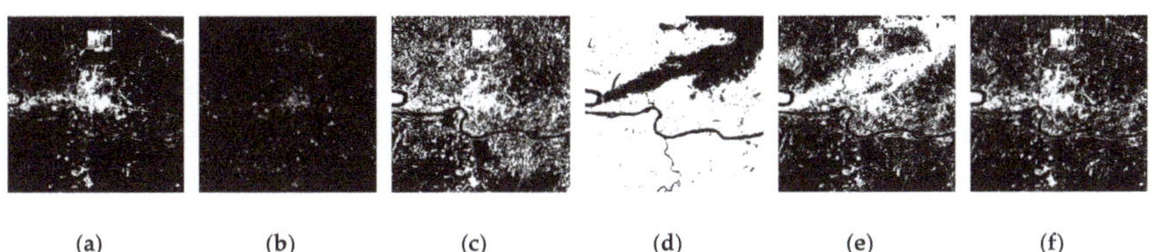

(a)    (b)    (c)    (d)    (e)    (f)

**Figure 5.** The impervious surface distribution in Case 2 extracted by six methods after water bodies were masked: (**a**) RISI; (**b**) NDBI; (**c**) IBI; (**d**) NDISI; (**e**) NDII; (**f**) CBI.

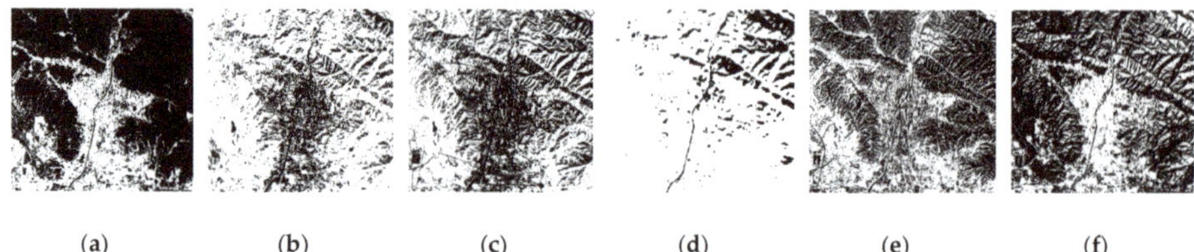

(a)          (b)          (c)          (d)          (e)          (f)

**Figure 6.** The impervious surface distribution in Case 3 extracted by six methods after water bodies were masked: (**a**) RISI; (**b**) NDBI; (**c**) IBI; (**d**) NDISI; (**e**) NDII; (**f**) CBI.

Figure 4 shows that the NDBI resulted in serious omissions while the other four reference methods provided higher accuracy. Figure 5 shows that the NDBI lost a large amount of impervious surfaces and other reference methods also resulted in serious errors. Figure 6 shows that all five reference methods mis-extracted a large amount of bare soil, and the NDBI and IBI lost a large amount impervious surfaces.

### 3.3. Accuracy

The recalls and precisions of the six extractions are shown in Figure 7, and the F1 scores are shown in Figure 8. The NDBI, IBI, NDISI, NDII, and CBI gave high recalls and low precisions, especially in Case 3 which had a large amount of bare soil, where the F1 scores were all less than 65%. The extraction accuracy by the proposed method was higher than others in our three case images, and the average values of the recall, precision, and F1 score were 95%, 91%, and 93%, respectively.

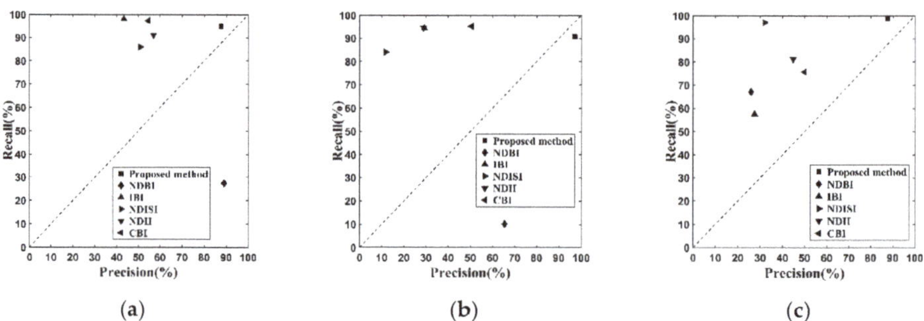

(a)                    (b)                    (c)

**Figure 7.** Recalls and precisions of the six extractions: (**a**) Case 1; (**b**) Case 2; (**c**) Case 3.

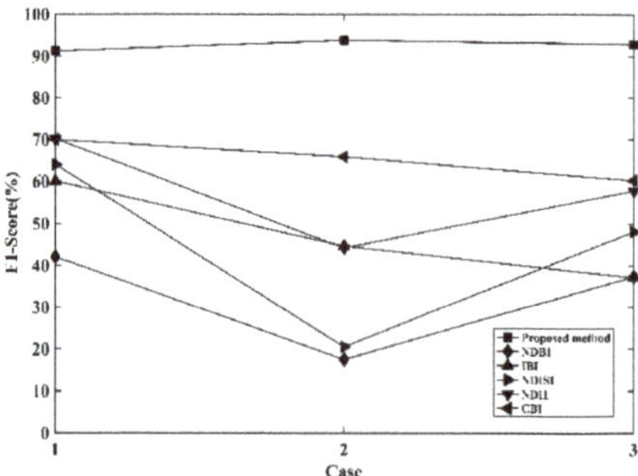

**Figure 8.** F1 scores of the six extractions.

## 4. Discussion

With the economic growth of developing countries, the rapid urbanisation of rural land and its conversion to urban land directly lead to an increase in the area of impervious surfaces, which may lead to the urban heat island effect. Impervious surfaces are artificial surfaces that water cannot infiltrate to reach the soil, such as parking lots, streets, and highways. They often have a low specific heat capacity and heat up fast, which often leads to the higher temperature in urban areas than rural areas, the uneven distribution of heat, the abnormal hydrothermal cycle, and the harmful urban heat island effect [35]. It is important to build accurate and fast methods to extract impervious surface distribution which is helpful for detecting regional environmental changes in urban areas and achieving sustainable urban development.

The existing methods for the extraction of impervious surface distribution increased the difference between impervious surfaces and other ground objects and achieved high precisions in some cases [7–11]. However, differences between sensors, regions, and seasons often lead to inconsistencies in the spectral features of ground objects in images. Many methods have poor abilities to distinguish impervious surfaces from bare soil because of their spectral similarity [36], and it is also not easy to distinguish withered vegetation in winter from impervious surfaces.

### 4.1. Ability of the Proposed Method to Distinguish Impervious Surfaces from Vegetation and Bare Soil

Figure 9 shows the differences between typical samples of impervious surfaces, bare soil, and vegetation in terms of the coastal band, blue band, green band, red band, ratio vegetation index (RVI) [37], NDVI, and difference vegetation index (DVI) [38], where all features underwent a 0–1 transformation. Of these three vegetation indices, the NDVI gave the greatest difference between impervious surfaces and other ground objects. Besides, the values for impervious surfaces in the coastal band were higher than those for vegetation and bare soil, and the contrast between impervious surfaces and bare soil gradually decreased in the coastal, blue, green, and red bands.

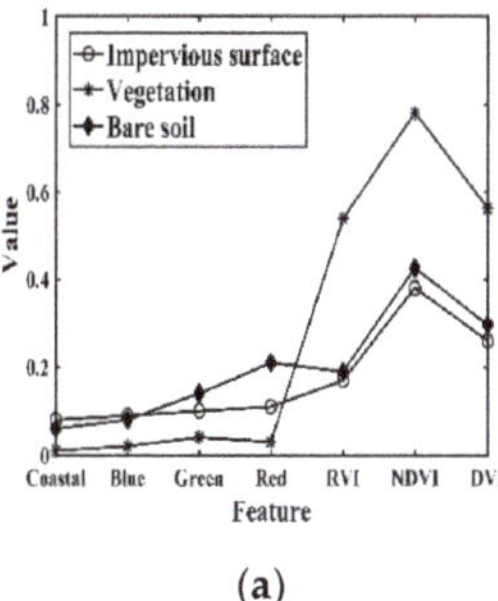

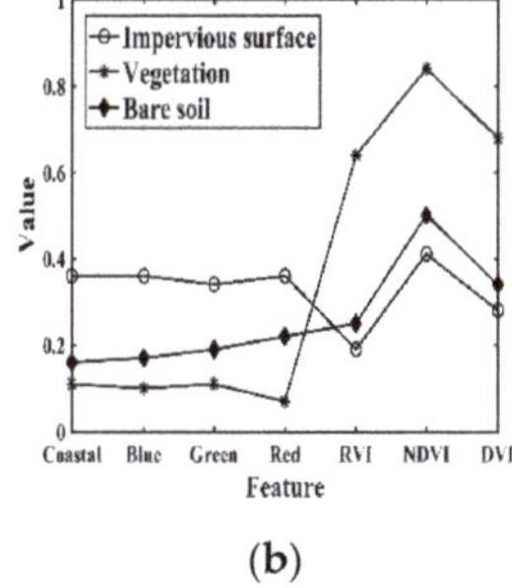

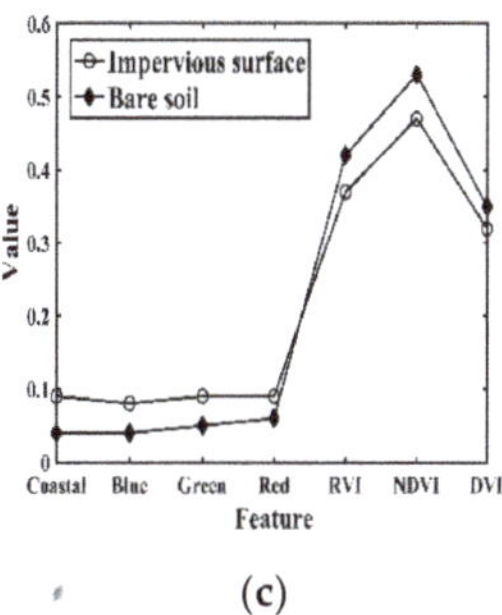


**Figure 9.** Ability of the proposed method to distinguish impervious surfaces from vegetation and bare soil: (**a**) Case 1; (**b**) Case 2; (**c**) Case 3.

Therefore, the ratio of the coastal band to the NDVI can maximise the information on impervious surfaces and increase the difference between impervious surfaces and other ground objects, and thus effectively distinguish impervious surfaces from bare soil, which is the main advantage of our method in comparison with the other five methods.

### 4.2. Extraction of Impervious Surface Distribution by Different Methods

The five methods used for comparison are applicable to different conditions and provide higher accuracy in other cases [7–11]. They exhibited differences in performance after water bodies were masked in our three cases, and the average F1 scores of extractions by six methods from the lowest to the highest value were in the following order: the NDBI, NDISI, IBI, NDII, CBI, and our proposed method.

The average F1 score of extractions by the NDBI after water bodies were masked was 32%. This index utilises the feature that the values for impervious surfaces in the SWIR1 band in Landsat 5 TM imagery are higher than those in the NIR band. However, the spectral features of impervious surfaces in Landsat 8 OLI imagery are different from those in Landsat 5 TM imagery because of the differences in the sensor, wavelength range, and quantisation bit. Figure 2 shows that the values for impervious surfaces in the SWIR1 band were comparable to those in the NIR band in Cases 1 and 3, whereas the values in the SWIR1 band were lower than those in the NIR band in Case 2. Conversely, the bare soil in Case 3 gave higher values in the SWIR1 band than in the NIR band, which resulted in more instances of error.

The average F1 score of extractions by the NDISI after water bodies were masked was 44%. The author who devised the NDISI pointed out that the values for impervious surfaces are higher than those of other ground objects in the thermal infrared band. This index led to instances of error in Cases 2 and 3, which were caused by the fact that the values for impervious surfaces in the Landsat 8 thermal infrared band were not always higher than those for other ground objects (Figure 2). Thermal radiation from impervious surfaces is more obvious in summer but may be lower than that from other ground objects in winter [39,40]. Therefore, the use of the NDISI for the extraction of impervious surface distribution in summer may provide high accuracy [9], but its accuracy may be low if it is employed in winter.

The average F1 score of extractions by the IBI after water bodies were masked was 47%. This index combines three indices, namely, the NDBI, MNDWI, and SAVI. In our cases, the NDBI values for impervious surfaces were not always higher than those for other ground objects such as the bare soil in Case 3 (Figure 10), which may have led to many instances of error.

The average F1 score of extractions by the NDII after water bodies were masked was 57%. Xu et al. pointed out that this index is unreliable because the values for impervious surfaces in the thermal infrared band are significantly lower than those in the visible light band [36]. Besides, this index requires that bare soil is masked in advance [10], which is difficult in practice.

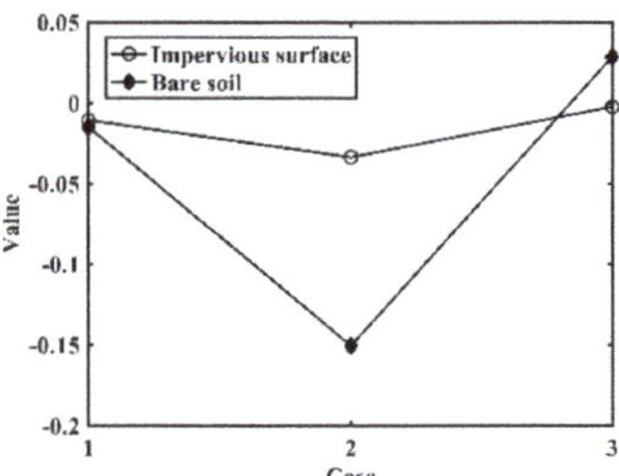


**Figure 10.** NDBI values for impervious surfaces and bare soil.

The average F1 score of extractions by the CBI after water bodies were masked was 65%. This index combines the PC1, NDWI, and SAVI, which represent high-albedo surfaces, low-albedo surfaces, and vegetation, respectively, but the ability to mask bare soil needs to be further enhanced [36].

The average F1 score of extractions by our proposed method was 93%. Our method not only identified impervious surfaces but also masked other ground objects such as bare soil and therefore provided higher accuracy than the other five methods.

*4.3. Atmospheric Effect of the Proposed Method*

The abovementioned impervious surface distributions were extracted on the basis of the DN values in the images.

When electromagnetic waves are transmitted across the atmosphere, they are scattered because of molecules and minute particles; the shorter the wavelength is, the stronger the scattering is, which causes radiation distortion in the coastal and blue bands of Landsat 8 images. Atmospheric correction of images is usually required in order to obtain more realistic surface reflectivity, but this is often a cumbersome process.

As a comparison, the case images were corrected according to the FLAASH model in ENVI, and reflectance values that were less than 0 or greater than 100 were modified according to Equation (16):

$$g(x,y) = \begin{cases} 0 & f(x,y) < 0 \\ f(x,y) & 0 \le f(x,y) \le 100 \\ 100 & f(x,y) > 100 \end{cases} \tag{16}$$

where $f(x,y)$ and $g(x,y)$ represent the images before and after modification, respectively.

The impervious surface distributions extracted after atmospheric correction are shown in Figure 11. From a comparison with Table 1, impervious surfaces were less widely distributed, and the average recall, precision, and F1 score were 89%, 77%, and 83%, respectively, which represented decreases by 6%, 14%, and 10%, respectively.

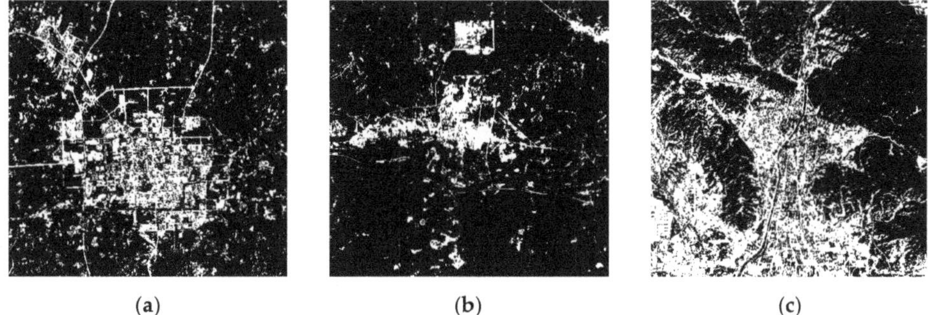

(a)        (b)        (c)

**Figure 11.** The impervious surface distribution extracted by the proposed method after atmospheric correction: (**a**) Case 1; (**b**) Case 2; (**c**) Case 3.

Atmospheric correction did not provide additional information that could supplement the results based on the DN, but instead resulted in more mis-extraction and losing results.

Taking Case 3 as an example, Figure 12 shows the difference between the impervious surface distribution extracted before and after atmospheric correction, where (a) is a false colour composite of the (5, 4, 3) band and (b) is an RGB composite of the impervious surface distribution extracted before atmospheric correction, ground truth for the impervious surface distribution and the impervious surface distribution extracted after atmospheric correction. Here, the white areas were extracted in both instances, the green areas were lost in both instances, the purple areas were mis-extracted in both instances, the yellow areas were lost after atmospheric correction, and the blue areas were only mis-extracted after atmospheric correction.

Actually, atmospheric correction using an atmospheric correction model is not always necessary and may decrease the accuracy of the extraction of remote sensing information [41]. With respect to our three case images, atmospheric correction did not help to improve the accuracy of the extraction of impervious surface distribution.

(a)

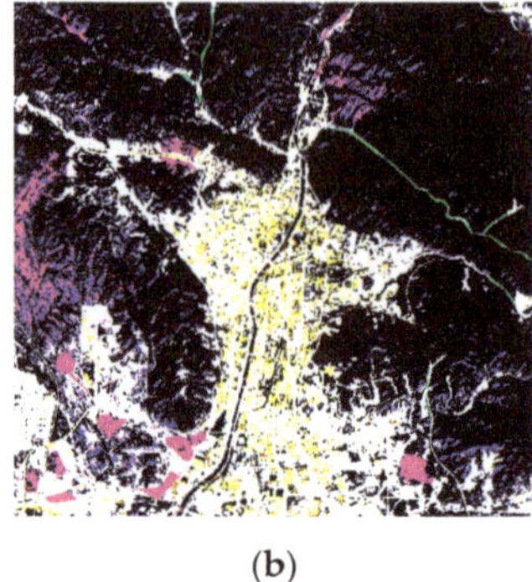

(b)

**Figure 12.** The impervious surface distribution extracted from Case 3 image before and after atmospheric correction: (**a**) (5, 4, 3) false colour composite with a linear stretch of 2%; (**b**) RGB composite of the impervious surface distribution extracted before atmospheric correction, ground truth for the impervious surface distribution and the impervious surface distribution extracted after atmospheric correction.

### 4.4. Alternatives to the Coastal Band

There is no coastal band in other Landsat imagery, which may influence the applicability of the proposed method. Here, we compare and discuss the differences that occur if impervious surface distribution is extracted using the blue band instead of the coastal band.

The differences between the values for typical samples of impervious surfaces and other ground objects of the ratio of the coastal band to the NDVI (coastal band/NDVI) and the ratio of the blue band to the NDVI (blue band/NDVI) were calculated, as shown in Figure 13. We can see that if the blue band/NDVI is used, the difference between impervious surfaces and other ground objects decreases slightly.

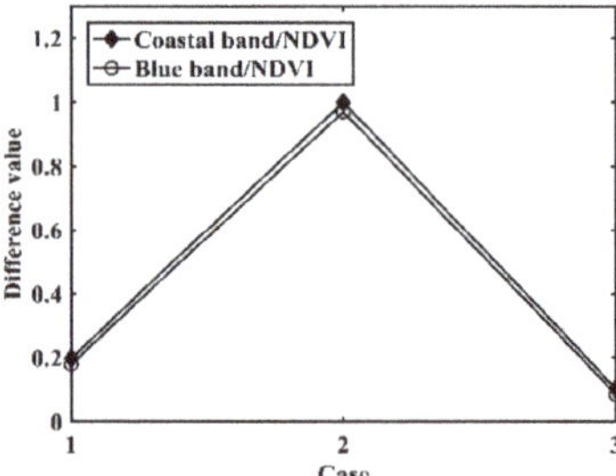



**Figure 13.** Differences in the coastal band/NDVI and blue band/NDVI between impervious surfaces and other ground objects.

The coastal band was replaced with the blue band, and the other parameters were the same as those used in the previous workflow to extract the impervious surface distribution from the above three case images. We also determined the accuracy and found that the average recall, precision, and F1 score were 93%, 87%, and 90%, respectively, which represented decreases by 2%, 4%, and 3%, respectively.

Therefore, it is feasible to use the blue band to extract impervious surface distribution from other Landsat remote sensing images without using the coastal band, albeit with a certain decrease in accuracy.

In contrast to validation using images for a single region and season [7–10], our study used three images obtained in summer and winter. The RISI exhibited higher accuracy and better applicability than the other methods used for comparison, and it avoids the additional feature selection and classifier selection processes used in classification methods.

## 5. Conclusions

This study proposed a novel index, namely, the RISI, for extracting impervious surface distribution from Landsat 8 imagery and validated it using three images covering urban areas of China in winter and summer. The results of the extraction were compared with those obtained using the NDBI, IBI, NDISI, NDII, and CBI. The main conclusions are as follows:

(1) The RISI effectively distinguishes impervious surfaces from other ground objects, especially bare soil, and gives higher recalls, precisions, and F1 scores, of which the average values for the three case images were 95%, 91%, and 93%, respectively. The validation confirmed the robustness of the RISI.

(2) The use of the coastal band after a 0–1 transformation can further increase the difference between impervious surfaces and bare soil.

(3) The workflow has been shown to have good applicability in the extraction of impervious surface distribution from Landsat imagery.

It is still necessary to perform validation using more images with complex compositions and distributions of ground objects to refine the workflow so that it suits more situations.

**Author Contributions:** H.F. performed the experiments, analyzed the data, and wrote the paper; Y.W. revised and commented on the manuscript; Q.D. performed the data processing and examination of the experiments.

**Funding:** This study is supported by the National Natural Science Foundation of China (No. 41471283).

**Acknowledgments:** The authors would like to thank the reviewers and the editor for precise and excellent comments that greatly improved this paper.



## References

1. Weng, Q. Remote sensing of impervious surfaces in the urban areas: Requirements, methods, and trends. *Remote Sens. Environ.* **2012**, *117*, 34–49. [CrossRef]
2. Mckinney, M.L. Urbanization, biodiversity, and conservation the impacts of urbanization on native species are poorly studied, but educating a highly urbanized human population about these impacts can greatly improve species conservation in all ecosystems. *Bioscience* **2002**, *52*, 883–890. [CrossRef]
3. Jacobson, M.Z.; Ten Hoeve, J.E. Effects of urban surfaces and white roofs on global and regional climate. *J. Clim.* **2012**, *25*, 1028–1044. [CrossRef]
4. Huszar, P.; Halenka, T.; Belda, M.; Zak, M.; Sindelarova, K.; Miksovsky, J. Regional climate model assessment of the urban land-surface forcing over central europe. *Atmos. Chem. Phys.* **2014**, *14*, 18541–18589. [CrossRef]
5. Jian, Y.; Li, P. Impervious surface extraction in urban areas from high spatial resolution imagery using linear spectral unmixing. *Remote Sens. Appl. Soc. Environ.* **2015**, *1*, 61–71.
6. Gao, Z.; Liu, X. Support vector machine and object-oriented classification for urban impervious surface extraction from satellite imagery. In Proceedings of the International Conference on Agro-geoinformatics, Beijing, China, 11–14 August 2014.
7. Zha, Y.; Ni, S.X.; Yang, S. An effective approach to automatically extract urban land-use from tm lmagery. *J. Remote Sens.* **2003**, *7*, 37–41.
8. Xu, H.Q. A new index for delineating built-up land features in satellite imagery. *Int. J. Remote Sens.* **2008**, *29*, 4269–4276. [CrossRef]
9. Xu, H.Q. Analysis of impervious surface and its impact on urban heat environment using the normalized difference impervious surface index (ndisi). *Photogramm. Eng. Remote Sens.* **2010**, *76*, 557–565. [CrossRef]
10. Wang, Z.; Gang, C.; Li, X.; Chen, Y.; Li, J. Application of a normalized difference impervious index (ndii) to extract urban impervious surface features based on landsat tm images. *Int. J. Remote Sens.* **2015**, *36*, 1055–1069. [CrossRef]
11. Sun, G.Y.; Chen, X.L.; Jia, X.P.; Yao, Y.J.; Wang, Z.J. Combinational build-up index (cbi) for effective impervious surface mapping in urban areas. *IEEE J. Sel. Top. Appl. Earth Obs. Remote Sens.* **2017**, *9*, 2081–2092. [CrossRef]
12. Voorde, T.V.D.; Roeck, T.D.; Canters, F. A comparison of two spectral mixture modelling approaches for impervious surface mapping in urban areas. *Int. J. Remote Sens.* **2009**, *30*, 4785–4806. [CrossRef]

13. Zhao, Y.; Xu, J. Impervious surface extraction with linear spectral mixture analysis integrating principal components analysis and normalized difference building index. In Proceedings of the International Workshop on Earth Observation and Remote Sensing Applications, Guangzhou, China, 4–6 July 2016.

14. Roberts, D.A.; Gardner, M.; Church, R.; Ustin, S.; Scheer, G.; Green, R.O. Mapping chaparral in the santa monica mountains using multiple endmember spectral mixture models. *Remote Sens. Environ.* **1998**, *65*, 267–279. [CrossRef]

15. Fan, F.; Deng, Y. Enhancing endmember selection in multiple endmember spectral mixture analysis (mesma) for urban impervious surface area mapping using spectral angle and spectral distance parameters. *Int. J. Appl. Earth Obs. Geoinf.* **2014**, *33*, 290–301. [CrossRef]

16. Cortes, C.; Vapnik, V. Support-vector networks. *Mach. Learn.* **1995**, *20*, 273–297. [CrossRef]

17. Dreiseitl, S.; Ohno-Machado, L. Logistic regression and artificial neural network classification models: A methodology review. *J. Biomed. Inform.* **2002**, *35*, 352–359. [CrossRef]

18. Breiman, L. Random forests. *Mach. Learn.* **2001**, *45*, 5–32. [CrossRef]

19. Tehrany, M.S.; Pradhan, B.; Jebuv, M.N. A comparative assessment between object and pixel-based classification approaches for land use/land cover mapping using spot 5 imagery. *Geocarto Int.* **2014**, *29*, 351–369. [CrossRef]

20. Patel, N.; Mukherjee, R. Extraction of impervious features from spectral indices using artificial neural network. *Arab. J. Geosci.* **2015**, *8*, 3729–3741. [CrossRef]

21. Pelletier, C.; Valero, S.; Inglada, J.; Champion, N.; Dedieu, G. Assessing the robustness of random forests to map land cover with high resolution satellite image time series over large areas. *Remote Sens. Environ.* **2016**, *187*, 156–168. [CrossRef]

22. Bhatti, S.S.; Tripathi, N.K. Built-up area extraction using landsat 8 oli imagery. *Gisci. Remote Sens.* **2014**, *51*, 445–467. [CrossRef]

23. Geospatial Data Cloud. Available online: http://www.gscloud.cn (accessed on 10 January 2019).

24. Ridd, M.K. Exploring a v-i-s (vegetation-impervious surface-soil) model for urban ecosystem analysis through remote sensing: Comparative anatomy for cities. *Int. J. Remote Sens.* **1995**, *16*, 2165–2185. [CrossRef]

25. Thanapura, P.; Helder, D.L.; Burckhard, S.; Warmath, E.; O'Neill, M.; Galster, D. Mapping urban land cover using quickbird ndvi and gis spatial modeling for runoff coefficient determination. *Photogramm. Eng. Remote Sens.* **2015**, *73*, 57–65. [CrossRef]

26. Mu, Y.C.; Jie, Y.W.; Zhang, L.L. An enhanced normalized difference impervious surface index. *Sci. Surv. Mapp.* **2018**, *43*, 83–87.

27. Chen, Q.; Li, X.T. Effects of New Characteristics of Landsat 8 Operational Land Imager (OLI) Data on Land-cover Remote Sensing Classification. *J. Subtrop. Resour. Environ.* **2015**, *10*, 79–86.

28. Ohtsu, N. A threshold selection method from gray-level histograms. *IEEE Trans. Syst. Man Cybern.* **1979**, *9*, 62–66. [CrossRef]

29. Mapper, E. *Er Mapper 6.0 User Guide*; 922 oldal; Earth Resource Mapping Pty Ltd.: West Perth, Australia, 1998.

30. Rouse, J.W.J.; Haas, R.H.; Schell, J.A.; Deering, D.W. Monitoring vegetation systems in the great plains with erts. *Nasa Spec. Publ.* **1973**, *351*, 309–317.

31. Huete, A.R. A soil-adjusted vegetation index (savi). *Remote Sens. Environ.* **1988**, *25*, 295–309. [CrossRef]

32. Xu, H.Q. Modification of normalised difference water index (ndwi) to enhance open water features in remotely sensed imagery. *Int. J. Remote Sens.* **2006**, *27*, 3025–3033. [CrossRef]

33. Mcfeeters, S.K. The use of the normalized difference water index (ndwi) in the delineation of open water features. *Int. J. Remote Sens.* **1996**, *17*, 1425–1432. [CrossRef]

34. Zhou, Z.H. *Machine Learning*; Tsinghua University Press: Beijing, China, 2016.

35. Mohajerani, A.; Bakaric, J.; Jeffrey-Bailey, T.; Mohajerani, A.; Bakaric, J. The urban heat island effect, its causes, and mitigation, with reference to the thermal properties of asphalt concrete. *J. Environ. Manag.* **2017**, *197*, 522–538. [CrossRef]

36. Xu, H.Q.; Wang, M.Y. Remote sensing-based retrieval of ground impervious surfaces. *J. Remote Sens.* **2016**, *20*, 1270–1289.

37. Jordan, C.F. Derivation of leaf-area index from quality of light on the forest floor. *Ecology* **1969**, *50*, 663–666. [CrossRef]

38. Tucker, C.J. Red and photographic infrared linear combinations for monitoring vegetation. *Remote Sens. Environ.* **1979**, *8*, 127–150. [CrossRef]

39. Wang, F.; Lan, X.J.; Wu, Z.G.; Chen, S.B. Monitoring analysis of heat island effects in Changshu city based on Landsat8. *J. Jiangxi Univ. Sci. Technol.* **2016**, *37*, 14–19.

40. Li, W.; Li, Q.X.; Jiang, Z.H. Analysis of the Influence of Urban Heat Island Effect on Temperature Warming in Southeastern China. In Proceedings of the Annual Meeting of the Chinese Meteorological Society, Hangzhou, China, 13–16 October 2009.

41. Song, C.; Woodcock, C.E.; Seto, K.C.; Lenney, M.P.; Macomber, S.A. Classification and change detection using landsat tm data: When and how to correct atmospheric effects. *Remote Sens. Environ.* **2001**, *75*, 230–244. [CrossRef]

*Article*

# A Hybrid Approach Using Fuzzy AHP-TOPSIS Assessing Environmental Conflicts in the Titan Mining Industry along Central Coast Vietnam

Manh Tien Dao [1], An Thinh Nguyen [2,*], The Kien Nguyen [2], Ha T.T. Pham [3], Dinh Tien Nguyen [2], Quoc Toan Tran [2], Huong Giang Dao [1], Duyen T. Nguyen [1], Huong T. Dang [1] and Luc Hens [4]

1   Institute of Resource, Environment and Sustainable Development, Hanoi 10000, Vietnam
2   Faculty of Development Economics, University of Economics and Business, Vietnam National University (VNU), Hanoi 10000, Vietnam
3   Faculty of Environmental Sciences, University of Sciences, Vietnam National University (VNU), Hanoi 10000, Vietnam
4   Vlaamse Instelling voor Technologisch Onderzoek (VITO), Boeretang 200, 2400 Mol, Belgium
*   Correspondence: nathinh@vnu.edu.vn; Tel.: +84-912-300-314

Received: 1 July 2019; Accepted: 18 July 2019; Published: 22 July 2019

**Abstract:** Environmental conflict management gains significance in rational use of natural resources, ecosystem preservation and environmental planning for mineral mines. In Central Coast Vietnam, titan mines are subject to conflicting use and management decisions. The paper deals with an empirical research on applying a combination of the fuzzy Analytic Hierarchy Process (AHP) and the fuzzy Technique for Order of Preference by Similarity to Ideal Solution (TOPSIS) to measure environmental conflicts emerging as a result of titan mining in Vietnam. The methodology used in the paper combines the fuzzy AHP and the fuzzy TOPSIS to rank environmental conflicts and propose conflict prevention solutions in the titan mining industry of Ky Khang coastal commune (Ky Anh district, Central Coast Vietnam). Data was collected by using a questionnaire with 15 locals, 8 communal authorities, 2 district authorities, and 12 scientific experts on titan mining, environmental geology, and sustainability management. The result shows that, titan mining conflicts with the eight criteria of economic sectors at five alternative sites including beach, protected forest, agricultural area, settlement area, and industrial area. The conflicts between titan mining and forestry, agriculture, settlements, fishing and aquaculture are highly valued. The beach area shows most environmental conflict as a result of titan mining, followed by the agricultural area and settlement area. Based on the empirical findings, legal and procedural tools such as environmental impact assessments, strategic environmental assessments, integrated coastal zone management, marine spatial planning, and multi-planning integration advancing environmental management for titan mines in Vietnam are suggested.

**Keywords:** fuzzy AHP; fuzzy TOPSIS; titan mining; environmental conflict; Ky Anh; Vung Ang Economic Zone; Central Coast Vietnam

---

## 1. Introduction

Environmental conflicts, which originate as a result of environmental pollutions, resource use competition, and social conflicts, emerge when stakeholders take part in activities with contradictory interests, values, power, perceptions, and goals. Environmental conflicts cover different issues: Biodiversity conflicts [1–3], coastal zone conflicts [4–7], air pollution conflicts [8], land use conflicts [9,10], and water conflicts [11–14]. Most recently, environmental conflict is considered in relation to economic, development and social issues in the context of global climate change [15,16]. Worldwide, environmental conflicts challenge the economic security at both local, regional, national, and global scales [17,18]. Particularly in coastal zones, environmental conflicts occur as a result of the negative impacts of environmental pollution by sectors and activities. The most sensitive and conflictive sectors are mining, land use, shrimp farming, fossil fuels, biomass, and hydropower plants [18–20]. In the mineral mining industry, environmental conflicts result from inadequate public information, stereotypes in decision making, the potential of problems to disappear, technical solutions, and archaic techniques [21]. Effective tools such as integrated coastal zone management (ICZM), marine spatial planning (MSP), and multi-planning integration advancing coastal zone management are recommended to address environmental conflicts in coastal areas [22–24].

The potential of mathematical models in environmental conflict analysis is internationally recognized. The General Algebraic Modelling System (GAMS) is applied to analyze coastal land conflicts [25]. The Analytic Network Process (ANP) is combined with the Driver-Pressure-State-Impact-Response (DPSIR) model to address conflict [26]. Applying multiple-criteria decision-making (MCDM) to environmental decision-making allows defining optimal specifications to be applied to environmental conflicts. For example, the Analytic Hierarchy Process (AHP) is used to evaluate open pit coal production [27]. Intuitionistic fuzzy set (IF) and AHP is combined to select best drilling mud for drilling operations [28]. Fuzzy Technique for Order of Preference by Similarity to Ideal Solution (TOPSIS) is applied to select best compromise alternatives for water resource use [29]. AHP is integrated with Fuzzy TOPSIS to assess the conservation priority for six alternatives sites of a coastal area [30]. Integrating Delphi and Fuzzy AHP-TOPSIS allow weighting criteria and prioritizing heat stress indices in surface mining. In this case, Delphi extracts criteria based on the advantages of occupational health experts and selected criteria are weighed using the most suitable heat stress index based on Fuzzy TOPSIS [31].

This paper aims applying a hybrid approach using a combination of fuzzy AHP and fuzzy TOPSIS to measure environmental conflicts emerging as a result of titan mining along the Central Coast Vietnam. This area has an abundant potential for mineral mining of titanium and zircon [32]. Four key titan mining sites currently exist in this area: the Ky Khang mine (Ha Tinh province), Nhat Le (Quang Binh), De Gi (Binh Dinh), and Nhum (Binh Thuan) [33]. Titan mining contributes significantly to the local economy; however, it causes environmental problems. Most of titan mining sites are open pits with environmental pollution, degraded mangrove ecosystems, negatively affected human health and local livelihoods. Also, land use and land cover changes (LULCC) are evidenced: Vegetation cover is cutoff, and is replaced by temporary transportation infrastructure for large vehicles such as excavators, containers, and trucks. The environmental problems related to mining activities causing conflicts between stakeholders as well as conflicts between the titan mining industry with agriculture, forestry, infrastructure, and heritage sites are intense and result in major social conflicts.

The rest of the paper is organized as follows: Section 2 introduces a combination of the fuzzy AHP and fuzzy TOPSIS methodology; the results for a case analysis are indicated in Section 3 including ranking environmental conflicts and proposing conflict prevention solutions; finally, conclusion and recommendation are drawn in Section 4.

## 2. Methodology

### 2.1. Study Area

Ky Khang is one of seven coastal communes of the Ky Anh district (Ha Tinh province, Vietnam), located in the northeast of the district (Figure 1). The commune has a coastline of approximately 5 km, a total area of 26.3 square kilometers, and a total population of about 7000 people [34]. While Ky Khang contains one of four key titan mines along the Central Coast, the commune also is a key agricultural area in the Ha Tinh province with 728 hectares of agricultural land area [35]. The local economy mainly relies on vegetable and rice crop production, aquaculture, and titan mining. The titan mine covers 759 hectares. The mining permit is issued by the Ha Tinh Mineral and Trading Corporation since 1997 [33]. Ky Khang borders the Vung Ang Economic Zone (EZ), which is one of the seven key coastal economic zones prioritized by the Vietnamese Government for the period 2016–2020. Over 350 enterprises operate in the Vung Ang EZ, targeting steel smelting, thermal power plants, electricity generation, and deep-water port services. This industrial package puts fishermen's livelihood and sea water quality under pressure. An environmental incident rises from Vung Ang EZ in 2016, attracting both domestic and international attention, is the "Formosa environmental disaster". This problem came from a large source of toxic waste produced by the Formosa Ha Tinh Steel Corporation. The disaster, which spread over the coast of Central Vietnam including the Ha Tinh, Quang Binh, Quang Tri and Thua Thien-Hue provinces, killed sea fish, shrimp, clam, and coral reefs [36,37]. Moreover, in the Ky Anh district, Ky Khang is the commune most seriously damaged by natural disasters such as tropical storms and coastline erosion [35]. Combined natural disasters, environmental disasters and environmental pollutions raise environmental conflicts between the titan mining industry and the other economic sectors (agriculture, aquaculture, fishing, tourism, and forestry).

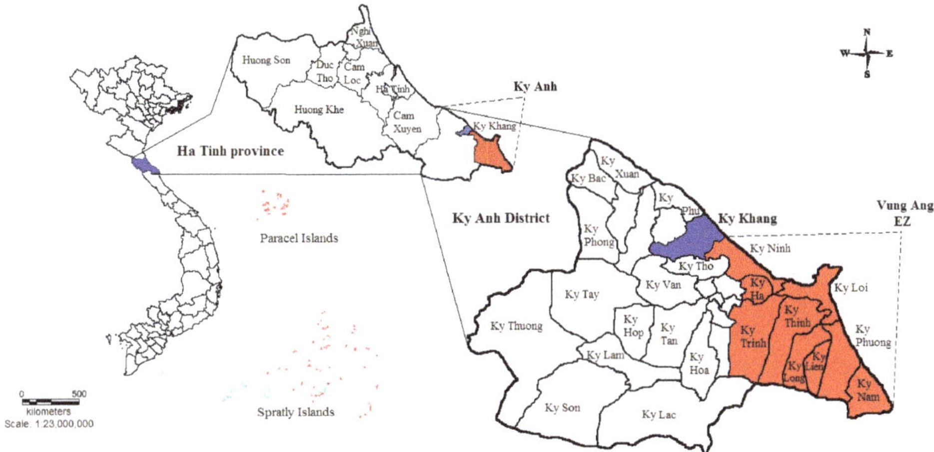


**Figure 1.** Location of Ky Khang and Vung Ang EZ in the Ky Anh district (Ha Tinh, Vietnam).

## 2.2. A Combination of Fuzzy AHP and Fuzzy TOPSIS

### 2.2.1. Fuzzy AHP

The AHP is one of the most common Multi-Criteria Decision Making (MCDM) instruments to deal with quantifiable and intangible criteria, which reflect the relative importance of the alternatives based on constructing a pair-wise comparison matrix [38–40]. Fuzzy AHP was developed to determine the weights of multiple criteria [41]. While the traditional AHP faces limitations on information imprecision and vagueness for decision making, the fuzzy AHP solves these problems of imprecision using linguistic variables, which are used to represent the relative importance between each pair of criteria [42]. Five linguistic variables corresponding with their fuzzy numbers are presented in Table 1.

**Table 1.** Linguistic variables and fuzzy numbers [42].

| Linguistic Variables | Fuzzy Numbers |
|---|---|
| Extreme importance (EXI) | $(7; 9; 9)$ |
| Very strong importance (VSI) | $(5; 7; 9)$ |
| Strong importance (SI) | $(3; 5; 7)$ |
| Moderate importance (MI) | $(1; 3; 5)$ |
| Equal importance (EI) | $(1; 1; 1)$ |

The process of Fuzzy AHP is structured in four steps [41]. Let $X = \{x_1, x_2 \dots x_n\}$ be an object set and $G = \{g_1, g_2 \dots g_m\}$ be a goal set. According to the extent analysis method [42], each object $x_i$ is evaluated by performing an extent analysis for each goal, $g_i$. Therefore, $m$ extent analysis values for each object can be obtained by using following notation: $M_{gi}{}^1, M_{gi}{}^2 \dots M_{gi}{}^m$.

Step 1. The fuzzy synthetic extent value $(S_i)$ with respect to the ith criterion is calculated in the following way:

$$S_i = \sum_{j=1}^{m} M_{gi}^{j} \otimes \left[ \sum_{i=1}^{n} \sum_{j=1}^{m} M_{gi}^{j} \right]^{-1} \tag{1}$$

With:

$$\sum_{j=1}^{m} M_{gi}^{j} = \left( \sum_{j=1}^{m} l_j, \sum_{j=1}^{m} m_j, \sum_{j=1}^{m} u_j \right), \sum_{i=1}^{n} \sum_{j=1}^{m} M_{gi}^{j} = \left( \sum_{i=1}^{n} l_i, \sum_{i=1}^{n} m_i, \sum_{i=1}^{n} u_i \right),$$

$$\left[ \sum_{i=1}^{n} \sum_{j=1}^{m} M_{gi}^{j} \right]^{-1} = \left( \frac{1}{\sum_{i=1}^{n} u_i}, \frac{1}{\sum_{i=1}^{n} m_i}, \frac{1}{\sum_{i=1}^{n} l_i} \right)$$

where: $l$ is the lower limit value; $m$ is the most promising value; $u$ is the upper limit value.

Step 2. The degree of possibility of $S_2(l_2, m_2, u_2) \geq S_1(l_1, m_1, u_1)$ is defined as:

$$V(S_1 \geq S_2) = \sup_{y \geq x} \left[ \min \left( \mu_{M_1}(x), \mu_{M_2}(y) \right) \right] \tag{2}$$

where: $x$ and $y$ are the values on the axis of membership function of each criterion.

$$V(S_1 \geq S_2) = \begin{cases} 1 \; if \; m_1 > m_2 \\ 0 \; if \; l_2 > u_1 \\ (l_2 - u_1)/(l_2 - u_1 + m_1 - m_2) \; otherwise \end{cases} \tag{3}$$

Step 3. The possibility that a convex fuzzy number $S$ is greater than $k$ convex fuzzy numbers $S_i \left( i = \overline{1,k} \right)$ is defined as:

$$V(S \geq S_1, S_2, \dots S_n) = V[(S \geq S_1), (S \geq S_2), \dots, (S \geq S_n)] = \min V(S \geq S_i);$$
$$\left( i = \overline{1,n} \right) \tag{4}$$

Step 4. Calculate the normalized weight vectors W′

$$W\prime = (d\prime(A_1), d\prime(A_2), \dots, d\prime(A_n))^T \tag{5}$$

where: $d\prime(A_i) = \min V(S_i \geq S_t); i = \overline{1,n}; t = \overline{1,n};$ and $i \neq t$

The normalized weight vectors $W'$ is generated according to the pairwise comparisons of the criteria of the involved respondents. As far as the important the corresponding criterion is concerned, the higher the weight, the more important the corresponding criterion.

### 2.2.2. Fuzzy TOPSIS

The fuzzy TOPSIS was developed based on the attribute of the shortest and the longest distance from the positive ideal solution and the negative ideal solution [43]. The best alternative has the shortest distance to the positive ideal solution, and the longest distance from the negative ideal solution. The classical TOPSIS assumes that individual preferences are assigned with crisp values. However, one should consider uncertainty and imprecision in the environmental practices. Fuzzy TOPSIS is more feasible because it incorporates the fuzzy environment uncertainty in decision making.

The process of Fuzzy TOPSIS used in this study follows six steps [43].

Step 1. Calculate the aggregate fuzzy ratings for the solutions.

If $x_{ijt} = \left(e_{ijt}, f_{ijt}, g_{ijt}\right)$, $i = \overline{1,n}$, $j = \overline{1,m}$, $t = \overline{1,l}$ is the fuzzy aggregated rating of solution $A_i$, by decision maker $D_t$, with respect to each criteria $C_j$. The fuzzy aggregated rating $x_{ij} = \left(e_{ij}, f_{ij}, g_{ij}\right)$, is given by:

$$x_{ij} = \frac{1}{l} \otimes \left(x_{ij1} \oplus x_{ij2} \oplus \ldots \oplus x_{ijt} \oplus \ldots \oplus x_{ijl}\right) \tag{6}$$

where: $e_{ij} = \frac{1}{l} \sum_{t=1}^{l} e_{ijt}$, $f_{ij} = \frac{1}{l} \sum_{t=1}^{l} f_{ijt}$, and $g_{ij} = \frac{1}{l} \sum_{t=1}^{l} g_{ijt}$

Step 2. Calculate normalized fuzzy decision matrix.

Data is normalized to obtain a comparable scale using linear scale transformation. Suppose $r_{ij} = \left(a_{ij}, b_{ij}, c_{ij}\right)$ is the mean of alternative solutions $i$ for criterion $j$. The normalized value $x_{ij}$ can be calculated as:

$$x_{ij} = \left(\frac{a_{ij}}{c_j^*}, \frac{b_{ij}}{c_j^*}, \frac{c_{ij}}{c_j^*}\right), j \in B \tag{7}$$

$$x_{ij} = \left(\frac{a_j^-}{c_{ij}}, \frac{a_j^-}{b_{ij}}, \frac{a_j^-}{a_{ij}}\right), j \in C \tag{8}$$

where: $a_j^- = \min_i a_{ij}$, $c_j^* = \max_i c_{ij}$, $i = \overline{1,n}$, $j = \overline{1,m}$

Step 3. Construct the weighted normalized matrix.

The weighted-normalized value $G_i$ is given by multiplying the normalized value $x_{ij}$ of the decision matrix by the weight assigned to the criterion $j$.

$$G_i = x_{ij} \otimes w_j, \ i = \overline{1,n}, \ j = \overline{1,m} \tag{9}$$

Step 4. Determine the fuzzy positive ideal and negative ideal solutions.

The positive ideal solution maximizes the benefit criteria and minimizes the cost criteria; whereas the negative ideal solution maximizes the cost criteria and minimizes the benefit criteria. The positive and negative ideal solutions are found out in the following way:

$$A^+ = \left[A_j^+\right]_{1j} = \left[v_{jmax}\right]_{1j} \tag{10}$$

$$A^- = \left[A_j^-\right]_{1j} = \left[v_{jmin}\right]_{1j}, \ j = \overline{1,m} \tag{11}$$

Calculate the distance/separation from:
* Positive Ideal Separation ($d^+$):

$$d_i^+ = \sqrt{\sum_{i=1}^{n} (G_i - A^+)^2} \tag{12}$$

* Negative Ideal Separation ($d^-$):

$$d_i^- = \sqrt{\sum_{i=1}^{n} (G_i - A^-)^2} \tag{13}$$

where: $d_i^+$ and $d_i^-$ are the distances of each alternative $A_i$ from positive and negative ideal solutions.

Step 5. Calculate the Relative closeness coefficient (CC) to the ideal solution.

$$CC_i = \frac{d_i^-}{d_i^+ + d_i^-}; \ 0 < CC_i < 1; \ i = \overline{1,n} \tag{14}$$

$CC_i = 1$ if $A_i = A^+$; $CC_i = 0$ if $A_i = A^-$

Step 6. Rank alternatives

The rank of considered alternatives can be decided, according to the descending order of closeness coefficient (CC). When the closeness coefficient is closer to 1 the corresponding solution is the optimal one.

### 2.2.3. Data Collection

The hybrid approach using Fuzzy AHP-TOPSIS provides a quantitative logical and systematic framework to identify critical issues, attach relative priorities to those issues, choose best compromise alternatives, and facilitate communication towards general acceptance [29,44]. This paper considered 8 criteria, representing the main economic sectors of Ky Khang, and 5 different alternative sites, representing conflict hotspots. Fuzzy AHP is combined with Fuzzy TOPSIS to estimate weights of the criteria and to prioritize alternative sites according to the intensity of environmental conflicts. Data on weighting sector criteria and prioritizing alternatives sites in the Ky Khang commune are collected using a questionnaire with 15 locals, 8 communal authorities, 2 district authorities, and 12 scientific experts on titan mining, environmental geology, and sustainability. Data were collected during a field trip in March 2018. All respondents inhabit in the study area (locals and authorities) or are knowledgeable about the study area and about scientific problems related to titan mining. The questionnaire allows inventory the opinion of the respondents on the pair wise comparison matrix by using Likert 5 scale, indicating five pre-coded responses with the neutral point (point 3) being neither agree nor disagree. Respondents took about 30 min to complete the questionnaire.

## 3. Results

### 3.1. Determining Criteria of Sectors and Alternative Sites

In the considered case study, in the area of Ky Khang, couples of environmental conflicts between the titan mining industry and other economic sectors are found across conflict hotspots. These couples are indicated by criteria in Fuzzy AHP model. Conflict hotspots are defined by alternative sites in Fuzzy TOPSIS model. It is supposed that conflicts occur based on eight criteria $(C_j)$ and are responsible for five alternatives sites $(A_i)$. The eight criteria entail agriculture (C1), aquaculture (C2), fishing (C3), salt production (C4), tourism (C5), forestry (C6), settlement (C7), and industry (C8) (Figure 2). Five alternatives sites include the beach (A1), a protected forest area (A2), an agricultural area (A3), a settlement area (A4), and an industrial area (A5). Since titan mining is popular along the coast, it conflicts with the eight criteria of sectors at the five alternatives sites. Impacts concern land use and competition with other sectors, use of fresh water during mining process, and emission of pollutants to air, soil and sediments. Consequently, mining destroys protection forests near the coast, reduces both surface and underground water quality and quantity, increases salinity, and destroys beaches.

In beach area (A1), titan mining impacts aquaculture, fishing, and tourism negatively by destroying beaches, eroding the coastline, and changing landscapes visually. Parts of titanium dumps were filled by clastic sedimentary rocks. Sand dunes have changed leaving hollow pits and deep holes. New sandy dunes of 6–10 m have appeared, consisting of silky sand, and being mobile by the wind. Mining leaves disposal sites and reservoirs over large areas.

Building infrastructure in a protected forest (A2) needs cutting trees, takes aquaculture ponds and mangrove farms, and kills sea fishes as a results of environmental pollution.

Titan mining takes arable land and decreases crop yields in agricultural area (A3). It causes competition spatially between titan mining and areas for salt production and decreases the quality and the yields of the salt due to pollutants. Crops on the fields and houses are flooded by sand. Mining damages soil surface: A large amount of soil and rocks are removed from the ground and leave holes behind. Surface water is polluted by overflowing acid effects, toxic pollutants and water-soluble solids release from deposits and sediments. Mine waste discharges sediments in rivers and streams.

Also, titan mining affects negatively buildings, heritage sites, and tourism infrastructures in the settlement area (A4). In industrial area (A5), titan mining and other industrial activities emit pollutants in the environment, which stress both locals and tourists. Mining machines, pumps and transport cause noise, which directly affects residents and tourists.

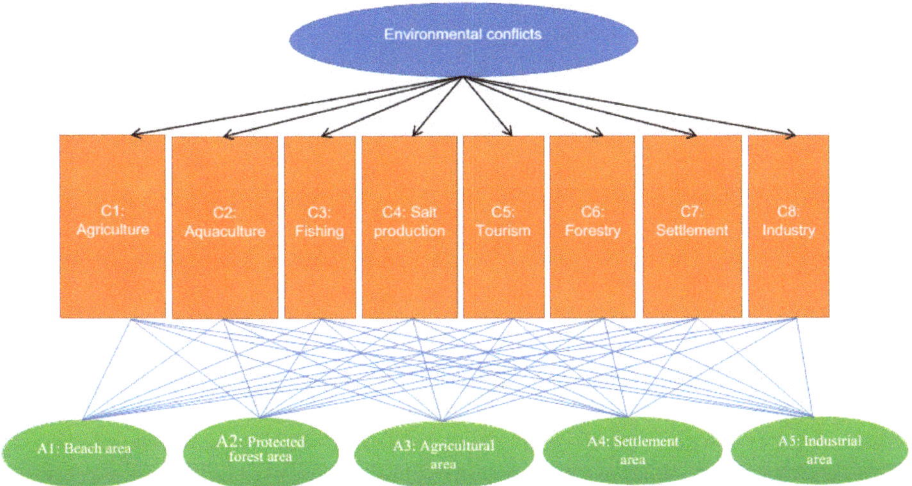


**Figure 2.** Decision tree on environmental conflicts with 8 criteria of sectors ($C_j$) in 5 alternative sites ($A_i$).

*3.2. Levels of Environmental Conflict and Priority Alternative Sites to Implement Conflict Prevention Solutions*

### 3.2.1. Weighting Criteria

The opinions of the different involved respondents (37 in total), collected by means of a questionnaire, were used to generate the pairwise comparisons of the criteria and determine the final importance levels represented in the decision matrices. The respondents determine the level of environmental conflicts between the titan mining industry with other economic sectors for five alternative sites. The evaluation of 5 decision matrices on 8 criteria is presented in Table 2. For example, comparing C1 and C2, C1 is considered of "very strong importance" compared to C2 for DM1 (value in the 4th column and in the 2nd raw). Consequently, the reciprocal value is assigned when C2 is compared to C1 for the same DM (3rd column, 7th raw).

Equations (1) and (2) are used to calculate the levels of the comparison between two fuzzy numbers based on the decision matrix. The relationship between a fuzzy number which is higher than the remaining fuzzy numbers, is calculated using Equations (3) and (4). Table 3 shows the results of priority weighting of the criteria based on the Equation (5). Forestry shows the highest score (0.131), which indicates titan mining has the most significant impacts on the forest plantations. Also, the conflicts between titan mining and agriculture, settlements, fishing and aquaculture are highly valued: agriculture shows the second weight score (0.129), followed by settlements (0.128), fishing (0.127), and aquaculture (0.125).

**Table 2.** Evaluation of decision matrix (DM).

| Criteria of Sectors | Decision Matrices | C1 | C2 | C3 | C4 | C5 | C6 | C7 | C8 |
|---|---|---|---|---|---|---|---|---|---|
| C1 (Agriculture) | DM1 | EI | VSI | EI | MI | MI | SI | EXI | SI |
| | DM2 | EI | SI | MI | VSI | VSI | (MI) | (SI) | SI |
| | DM3 | EI | SI | EI | SI | EXI | SI | SI | VSI |
| | DM4 | EI | (MI) | MI | SI | SI | (SI) | (SI) | (MI) |
| | DM5 | EI | MI | EI | SI | VSI | (SI) | (MI) | (MI) |
| C2 (Aquaculture) | DM1 | (VSI) | EI | (SI) | (MI) | EI | EI | VSI | MI |
| | DM2 | (SI) | EI | (MI) | SI | SI | (SI) | (VSI) | MI |
| | DM3 | (SI) | EI | (MI) | MI | EXI | EI | SI | VSI |
| | DM4 | MI | EI | SI | VSI | SI | (SI) | (SI) | (MI) |
| | DM5 | (MI) | EI | (MI) | MI | MI | (VSI) | (VSI) | (SI) |
| C3 (Fishing) | DM1 | SI | EI | EI | EI | MI | SI | EXI | VSI |
| | DM2 | MI | (MI) | EI | SI | VSI | (SI) | (SI) | SI |
| | DM3 | MI | EI | EI | SI | SI | MI | SI | VSI |
| | DM4 | (SI) | (MI) | EI | MI | EI | (VSI) | (SI) | (SI) |
| | DM5 | MI | EI | EI | MI | SI | (SI) | (SI) | (MI) |
| C4 (Salt production) | DM1 | MI | (MI) | EI | EI | MI | SI | EXI | VSI |
| | DM2 | (SI) | (VSI) | (SI) | EI | MI | (VSI) | (EXI) | (MI) |
| | DM3 | (MI) | (SI) | (SI) | EI | SI | (SI) | EI | MI |
| | DM4 | (VSI) | (SI) | (MI) | EI | (MI) | (EXI) | (VSI) | (SI) |
| | DM5 | (MI) | (SI) | (MI) | EI | MI | (EXI) | (VSI) | (SI) |
| C5 (Tourism) | DM1 | EI | (MI) | (MI) | (MI) | EI | MI | VSI | SI |
| | DM2 | (SI) | (VSI) | (VSI) | (MI) | EI | (EXI) | (EXI) | (SI) |
| | DM3 | (EXI) | (EXI) | (SI) | (SI) | EI | (VSI) | (SI) | (MI) |
| | DM4 | (SI) | (SI) | EI | MI | EI | (EXI) | (VSI) | (SI) |
| | DM5 | (MI) | (VSI) | (SI) | (MI) | EI | (EXI) | (EXI) | (VSI) |
| C6 (Forestry) | DM1 | EI | (SI) | (SI) | (SI) | (MI) | EI | SI | MI |
| | DM2 | SI | MI | SI | VSI | EXI | EI | (MI) | VSI |
| | DM3 | EI | (SI) | (MI) | SI | VSI | EI | SI | SI |
| | DM4 | SI | SI | VSI | EXI | EXI | EI | MI | MI |
| | DM5 | VSI | SI | SI | EXI | EXI | EI | MI | SI |
| C7 (Settlement) | DM1 | (VSI) | (EXI) | (EXI) | (EXI) | (VSI) | (SI) | EI | (MI) |
| | DM2 | VSI | SI | SI | EXI | EXI | MI | EI | EXI |
| | DM3 | (SI) | (SI) | (SI) | EI | SI | (SI) | EI | EI |
| | DM4 | SI | SI | SI | VSI | VSI | (MI) | EI | EI |
| | DM5 | VSI | MI | SI | VSI | EXI | (MI) | EI | EI |
| C8 (Industry) | DM1 | (MI) | (SI) | (VSI) | (VSI) | (SI) | (MI) | MI | EI |
| | DM2 | (MI) | (SI) | (SI) | MI | SI | (VSI) | (EXI) | EI |
| | DM3 | (VSI) | (VSI) | (VSI) | (MI) | MI | (SI) | EI | EI |
| | DM4 | MI | MI | SI | SI | SI | (MI) | EI | EI |
| | DM5 | SI | MI | MI | SI | VSI | (SI) | EI | EI |

Where: EXI = extreme importance; VSI = very strong importance; SI = strong importance; MI = moderate importance; EI = equal importance; the terms in brackets "()" indicate the reciprocal of them.

**Table 3.** Priority weighting matrix.

| V | S(C1) | S(C2) | S(C3) | S(C4) | S(C5) | S(C6) | S(C7) | S(C8) | d′ | Weight (w) | Ranking |
|---|---|---|---|---|---|---|---|---|---|---|---|
| S(C1)≥ | - | 0.97 | 0.99 | 1.00 | 1.00 | 0.96 | 0.98 | 1.00 | 0.96 | 0.129 | 2 |
| S(C2)≥ | 1.00 | - | 1.00 | 1.00 | 1.00 | 0.99 | 1.00 | 1.00 | 0.99 | 0.125 | 5 |
| S(C3)≥ | 1.00 | 0.98 | - | 1.00 | 1.00 | 0.97 | 0.99 | 1.00 | 0.97 | 0.127 | 4 |
| S(C4)≥ | 0.98 | 0.94 | 0.96 | - | 1.00 | 0.93 | 0.95 | 0.99 | 0.93 | 0.121 | 7 |
| S(C5)≥ | 0.95 | 0.91 | 0.93 | 0.98 | - | 0.89 | 0.92 | 0.96 | 0.89 | 0.116 | 8 |
| S(C6)≥ | 1.00 | 1.00 | 1.00 | 1.00 | 1.00 | - | 1.00 | 1.00 | 1.00 | 0.131 | 1 |
| S(C7)≥ | 1.00 | 0.99 | 1.00 | 1.00 | 1.00 | 0.98 | - | 1.00 | 0.98 | 0.128 | 3 |
| S(C8)≥ | 0.99 | 0.96 | 0.98 | 1.00 | 1.00 | 0.94 | 0.96 | - | 0.94 | 0.123 | 6 |

Where: C1 = Agriculture; C2 = Aquaculture; C3 = Fishing; C4 = Salt production; C5 = Tourism; C6 = Forestry; C7 = Settlement; C8 = Industry.

### 3.2.2. Final Ranking of the Alternatives

The opinions of the respondents are provided for eight criteria and potential options in five alternative sites. The linguistic variables are converted into triangular fuzzy numbers: S = (VL, L, M, H, VH), VL (Very low) = (0, 0, 0.2), L (Low) = (0, 0.2, 0.4), M (Medium) = (0.2, 0.4, 0.6), H (High) = (0.4, 0.6, 0.8), VH (Very high) = (0.6, 0.8, 1.0) (Table 4). Formula (6) allows evaluating the five decision matrices on the level of conflict. Formulas (7) and (8) are used to standardize the criteria. Formula (9) allows determining the value of the ratio weighted normalized and the distances of each alternative to the positive and negative ideal points. The values of separation from positive solution ($d^+$) and Separation from negative solution ($d^-$) are calculated by using Formulas (10)–(13). Values of closeness coefficient (CC) are calculated from values of $d^+$ and $d^-$ by means of formula (14). The weighted normalized decision matrices indicate the highest and the lowest weighted values belong to forestry and tourism respectively ($w_{(C6)} = 0.131$; $w_{(C5)} = 0.116$) (Table 5).

**Table 4.** Evaluation of decision matrix and average values of five alternative sites.

| Alternative Sites | Decision Matrices | C1 | C2 | C3 | C4 | C5 | C6 | C7 | C8 |
|---|---|---|---|---|---|---|---|---|---|
| A1 (Beach area) | DM1 | L | VH | H | H | M | M | L | VL |
| | DM2 | VL | VH | VH | VH | H | M | VL | L |
| | DM3 | VL | VH | VH | H | M | VH | VL | L |
| | DM4 | VL | L | L | M | VL | M | M | VL |
| | DM5 | M | VH | VH | VH | VH | H | M | H |
| A2 (Protected forest area) | DM1 | L | M | H | L | L | M | VL | VL |
| | DM2 | L | H | L | VL | H | VH | L | VL |
| | DM3 | L | VL | VL | L | M | VH | M | M |
| | DM4 | L | M | L | M | VL | VH | M | L |
| | DM5 | L | M | M | VL | H | VH | VL | VL |
| A3 (Agricultural area) | DM1 | M | M | L | M | VL | L | VL | L |
| | DM2 | VH | M | L | M | M | M | VH | L |
| | DM3 | VH | M | VL | L | VL | L | H | M |
| | DM4 | VH | M | M | H | L | M | L | L |
| | DM5 | VH | H | H | VL | M | L | L | L |
| A4 (Settlement area) | DM1 | VL | L | M | VL | L | L | H | L |
| | DM2 | M | VL | VL | M | M | H | VH | M |
| | DM3 | M | L | L | M | L | M | VH | M |
| | DM4 | M | L | M | L | L | L | VH | M |
| | DM5 | H | VL | VL | M | L | M | VH | M |
| A5 (Industrial area) | DM1 | M | M | VH | M | M | VH | VL | VL |
| | DM2 | VL | VL | VL | VL | VL | L | VH | VH |
| | DM3 | VL | VL | M | L | L | M | VH | VH |
| | DM4 | L | VL | M | M | L | VL | M | VH |
| | DM5 | VL | L | L | H | VL | VL | H | VH |

Where: DM1.5 is decision matrices; C1 = Agriculture; C2 = Aquaculture; C3 = Fishing; C4 = Salt production; C5 = Tourism; C6 = Forestry; C7 = Settlement; C8 = Industry.

**Table 5.** Weighted normalized decision matrices.

| Criteria of Sectors | A1 (Beach Area) | | | A2 (Protected Forest Area) | | | A3 (Agricultural Area) | | | A4 (Settlement Area) | | | A5 (Industrial Area) | | | Weight (w) |
|---|---|---|---|---|---|---|---|---|---|---|---|---|---|---|---|---|
| C1 | 0.04 | 0.13 | 0.35 | 0.00 | 0.22 | 0.43 | 0.57 | 0.78 | 1.00 | 0.22 | 0.39 | 0.61 | 0.04 | 0.13 | 0.35 | 0.129 |
| C2 | 0.55 | 0.77 | 1.00 | 0.23 | 0.41 | 0.64 | 0.27 | 0.50 | 0.73 | 0.00 | 0.14 | 0.36 | 0.05 | 0.14 | 0.36 | 0.125 |
| C3 | 0.52 | 0.76 | 1.00 | 0.14 | 0.33 | 0.57 | 0.14 | 0.33 | 0.57 | 0.10 | 0.24 | 0.48 | 0.24 | 0.43 | 0.67 | 0.127 |
| C4 | 0.52 | 0.76 | 1.00 | 0.05 | 0.19 | 0.43 | 0.19 | 0.38 | 0.62 | 0.14 | 0.33 | 0.57 | 0.19 | 0.38 | 0.62 | 0.121 |
| C5 | 0.44 | 0.69 | 1.00 | 0.31 | 0.56 | 0.88 | 0.13 | 0.31 | 0.63 | 0.06 | 0.38 | 0.69 | 0.06 | 0.25 | 0.56 | 0.116 |
| C6 | 0.35 | 0.57 | 0.78 | 0.57 | 0.78 | 1.00 | 0.09 | 0.30 | 0.52 | 0.17 | 0.39 | 0.61 | 0.17 | 0.30 | 0.52 | 0.131 |
| C7 | 0.08 | 0.21 | 0.42 | 0.08 | 0.21 | 0.42 | 0.21 | 0.38 | 0.58 | 0.58 | 0.79 | 1.00 | 0.38 | 0.54 | 0.75 | 0.128 |
| C8 | 0.09 | 0.22 | 0.43 | 0.04 | 0.13 | 0.35 | 0.04 | 0.26 | 0.48 | 0.17 | 0.39 | 0.61 | 0.52 | 0.70 | 0.91 | 0.123 |

The ideal solution comprises all of best values possible of criteria, whereas the negative ideal solution consists of all worst value possible of criteria. Calculated separation from negative ideal solution ($d^-$) indicates beach area (A1), agricultural area (A3) and settlement area (A4) have the highest scores (0.56, 0.46, and 0.44 respectively); whereas they get lowest scores for separation from positive ideal solution ($d^+$) (7.47, 7.58, and 7.60 respectively). According to closeness coefficient, the beach area (A1) shows most environmental conflict as a result of titan mining ($CC_{(A1)} = 0.0703$), followed by the agricultural area ($CC_{(A3)} = 0.0575$), and settlement area ($CC_{(A4)} = 0.0549$) (Table 6). Since the worst alternative sites have farthest distance from the ideal solution and the shortest distance from the negative ideal solution, it is necessary to pay more attention on environmental conflicts in these alternative sites.

**Table 6.** Final ranking alternatives.

**(a) Separation from Positive Ideal Solution ($d^+$)**

| Alternative Sites | Criteria of Sectors | | | | | | | | $d^+$ |
|---|---|---|---|---|---|---|---|---|---|
| | C1 | C2 | C3 | C4 | C5 | C6 | C7 | C8 | |
| A1 (Beach area) | 0.98 | 0.90 | 0.90 | 0.90 | 0.92 | 0.93 | 0.97 | 0.97 | 7.47 |
| A2 (Protected forest area) | 0.97 | 0.95 | 0.96 | 0.97 | 0.93 | 0.90 | 0.97 | 0.98 | 7.62 |
| A3 (Agricultural area) | 0.90 | 0.94 | 0.96 | 0.95 | 0.96 | 0.96 | 0.95 | 0.97 | 7.58 |
| A4 (Settlement area) | 0.95 | 0.98 | 0.97 | 0.96 | 0.96 | 0.95 | 0.90 | 0.95 | 7.60 |
| A5 (Industrial area) | 0.98 | 0.98 | 0.94 | 0.95 | 0.97 | 0.96 | 0.93 | 0.91 | 7.61 |

**(b) Separation from Negative Ideal Solution ($d^-$)**

| Alternative Sites | Criteria of Sectors | | | | | | | | $d^-$ |
|---|---|---|---|---|---|---|---|---|---|
| | C1 | C2 | C3 | C4 | C5 | C6 | C7 | C8 | |
| A1 (Beach area) | 0.03 | 0.10 | 0.10 | 0.10 | 0.09 | 0.08 | 0.04 | 0.04 | 0.56 |
| A2 (Protected forest area) | 0.04 | 0.06 | 0.05 | 0.03 | 0.07 | 0.10 | 0.04 | 0.03 | 0.42 |
| A3 (Agricultural area) | 0.10 | 0.07 | 0.05 | 0.05 | 0.05 | 0.05 | 0.05 | 0.04 | 0.46 |
| A4 (Settlement area) | 0.05 | 0.03 | 0.04 | 0.05 | 0.05 | 0.06 | 0.10 | 0.06 | 0.44 |
| A5 (Industrial area) | 0.03 | 0.03 | 0.06 | 0.05 | 0.04 | 0.05 | 0.07 | 0.10 | 0.43 |

Where: C1 = Agriculture; C2 = Aquaculture; C3 = Fishing; C4 = Salt production; C5 = Tourism; C6 = Forestry; C7 = Settlement; C8 = Industry; Closeness coefficient (CC): $CC_{(A1)}$ = 0.0703; $CC_{(A2)}$ = 0.0521; $CC_{(A3)}$ = 0.0575; $CC_{(A4)}$ = 0.0549; $CC_{(A5)}$ = 0.0534; Final ranking of alternatives: Level 1 (A1); level 2 (A3); level 3 (A4); level 4 (A5); level 5 (A2).

## 4. Conclusions

The paper presents empirical research on environmental conflicts emerging as a result of recent development of titan mining industry along the Central Coast Vietnam. The considered Ky Khang coastal area is located in an economically fast developing area. The resources are subject to conflicting use and management decisions. Coastal zone conflicts are fueled by a combination of other conflicts, which demonstrates the unique character of environmental conflicts along coast. Titan mining is a major problem resulting in most serious impacts on the ecosystems and the environment along coasts. Mining destroys protected forest areas, erodes coastlines, affects the ground water level, increases salinity and modifies natural landscapes. The risk of environmental conflicts between socio-economic groups along the coast increases during the mineral exploitation period.

A combination of Fuzzy AHP and Fuzzy TOPSIS shows its advantages in connecting decision makers with conflicting objectives to reach consensus. The systematic and logical approach is a strength since it allows solving complex multi-person and multi-criteria decision problems by weighting environmental conflicts and relating them to alternative sites. The model contributes to the group decision-making process taking into account the affected sites surrounding the titan mines. Determining the intensity of the conflicts and the alternative sites in which titan mining occurs is imperative. According to the results of this study, titan mining affects most seriously the forests. Since only few forests are left along the coast, environmental conflicts are obvious for protected forests. The raising of beach tourism, agricultural production, and housing all demand more space and compete with titan mining. Therefore, it is necessary to pay more attention to solutions mitigating conflicts in these sites.

To the best of our knowledge, till now no previous study has identified the ranks of environmental conflicts and the priorities of conflict prevention solutions in the case of coastal areas of Vietnam. Therefore, the paper makes an attempt to discuss management implications for environmental conflicts in mineral mines for Vietnam. These environmental conflicts are at risk of escalation and intensification in case they are not managed well [44–46]. The findings of this study suggested decisions should use decision science in identifying priority of prevention solutions for each environmental conflict hotspot. In the titan mine of studied Ky Khang, key issues of environmental conflicts concern climate change, biodiversity, environmental air quality, forestry, fresh water, and land resources. Environmental conflicts raise in Ky Khang and in the Vung Ang EZ because of the limited land areas and freshwater resources. Surface water is salinized by the mine activities and useless for agri- and aquaculture. Air is polluted by smoke, toxic gases and dust from the mining activities and electricity generation. While agriculture, forestry, aquaculture, salt production, tourism, heritage preservation, and human health all are affected by titan mining, the couple of conflicts titan mining—forestry should be considered most carefully in the study area. Managing environmental conflicts pertains to the available tools and approaches, and recommendations arising from these study results provide a scientific basis for policy-makers to mitigate the negative impacts of the conflicts. Since environmental conflicts dovetail in political, economic, social and ecological contexts, their management should focus on integrated measures [47,48]. Environmental impact assessments (EIA) and strategic environmental assessments (SEA) are important legal and procedural tools to manage the environment, in particular with regard to the current and intended development [49,50]. Moreover, integrated coastal zone management (ICZM), marine spatial planning (MSP), and multi-planning integration advancing coastal zone management are favorable to titan mines along coasts. In reality, there is lack of planning for natural resource use in the Central Coast Vietnam [35,51]. Consequently, resources in this area are depleted and risks of conflicts between economic groups using these natural resources are common. Over-all, conflicts will likely become more severe and serious as long as resources decline both in quality and quantity and environmental pollution increases. Developing natural resources and environmental management strategies and models of sustainable consumption of resources according to a sustainable development strategy are required. Potential conflicts are identified, and preventive measures are proposed.

**Author Contributions:** Conceptualization, M.T.D., A.T.N. and L.H.; Data Curation, Q.T.T.; Formal Analysis, D.T.N. (Dinh Tien Nguyen); Investigation, Q.T.T., H.G.D., D.T.N. (Duyen T. Nguyen) and H.T.D.; Methodology, A.T.N., T.K.N. and H.T.T.P.; Project Administration, M.T.D., H.G.D. and D.T.N. (Duyen T. Nguyen); Resources, Q.T.T. and D.T.N. (Duyen T. Nguyen); Software, T.K.N. and H.G.D.; Validation, D.T.N. (Dinh Tien Nguyen); Visualization, H.T.D.; Writing—Original Draft, M.T.D., A.T.N., H.T.T.P., D.T.N. (Dinh Tien Nguyen) and L.H.; Writing—Review & Editing, A.T.N. and L.H.

**Funding:** This research was funded by the Vietnam national project code DTDLCN.31/16.

**Acknowledgments:** The authors would like to thank locals, district and communal authorities, and scientists who were most collaborative in completing the questionnaire, and in providing discussion opportunities.

**Conflicts of Interest:** The authors declare no conflict of interest.

## References

1. Lu, Y.; Fu, B.; Chen, L.; Ouyang, Z.; Xu, J. Resolving the conflicts between biodiversity conservation and socioeconomic development in China: Fuzzy clustering approach. *Biodivers. Conserv.* **2006**, *15*, 2813–2827. [CrossRef]

2. Young, J.C.; Marzano, M.; White, R.M.; McCracken, I.; Redpath, S.M.; Carss, D.N.; Quine, C.P.; Watt, A.D. The emergence of biodiversity conflicts from biodiversity impacts: Characteristics and management strategies. *Biodivers. Conserv.* **2010**, *19*, 3973–3990. [CrossRef]

3. Meng, H.H.; Zhou, S.S.; Li, L.; Tan, Y.H.; Li, J.W.; Li, J. Conflict between biodiversity conservation and economic growth: Insight into rare plants in tropical China. *Biodivers. Conserv.* **2019**, *28*, 523–537. [CrossRef]

4. Cadoret, A. Conflict dynamics in coastal zones: A perspective using the example of Languedoc-Rousillon (France). *Coast. Conserv.* **2009**, *13*, 151. [CrossRef]

5. Bramatia, M.C.; Musellab, F.; Allevaa, G. What drives environmental conflicts in coastal areas? An econometric approach. *Ocean Coast. Manag.* **2014**, *101*, 63–78. [CrossRef]

6. Stepanova, O. Conflict resolution in coastal resource management: Comparative analysis of case studies from four European countries. *Ocean Coast. Manag.* **2015**, *103*, 109–122. [CrossRef]

7. Rempis, N.; Alexandrakis, G.; Tsilimigkas, G.; Kampanis, N. Coastal use synergies and conflicts evaluation in the framework of spatial, development and sectoral policies. *Ocean Coast. Manag.* **2018**, *166*, 40–51. [CrossRef]

8. Solomon, N.; Birhane, E.; Gordon, C.; Haile, M.; Taheri, F.; Azadi, H.; Scheffran, J. Environmental impacts and causes of conflict in the Horn of Africa: A review. *Earth-Sci. Rev.* **2018**, *177*, 284–290. [CrossRef]

9.  Barry, M.B.; Dewar, D.; Whittal, J.F.; Muzondo, I.F. Land Conflicts in Informal Settlements: Wallacedene in Cape Town, South Africa. *Urban Forum* **2007**, *18*, 171–189. [CrossRef]

10. Pape, R.; Löffler, J. Climate change, land use conflicts, predation and ecological degradation as challenges for reindeer husbandry in Northern Europe: What do we really know after half a century of research? *AMBIO* **2012**, *41*, 421–434. [CrossRef]

11. Zeitoun, M.; Mirumachi, N. Transboundary water interaction I: Reconsidering conflict and cooperation. *Int. Environ. Agreem. Politics Law Econ.* **2008**, *8*, 297. [CrossRef]

12. Strauß, S. Water conflicts among different user groups in South Bali, Indonesia. *Hum. Ecol.* **2011**, *39*, 69–79. [CrossRef]

13. Gunasekara, N.K.; Kazama, S.; Yamazaki, D.; Oki, T. Water conflict risk due to water resource availability and unequal distribution. *Water Resour. Manag.* **2014**, *28*, 169–184. [CrossRef]

14. Yuling, S.; Lein, H.; Chen, X. Water conflicts in Hetian District, Xinjiang, during the Republic of China period. *Water Hist.* **2016**, *8*, 77–94. [CrossRef]

15. Abrahams, D.; Carr, E.R. Understanding the connections between climate change and conflict: Contributions from geography and political ecology. *Curr. Clim. Chang. Rep.* **2017**, *3*, 233–242. [CrossRef]

16. Pearson, D.; Newman, P. Climate security and a vulnerability model for conflict prevention: A systematic literature review focusing on African agriculture. *Sustain. Earth* **2019**, *2*, 2. [CrossRef]

17. Brock, L. Environmental Conflict Research—Paradigms and Perspectives. In *Environmental Change and Security: International and European Environmental Policy Series*; Carius, A., Lietzmann, K.M., Eds.; Springer: Berlin/Heidelberg, Germany, 1999; pp. 37–53.

18. Klaus, R. Agrarian development and conflicts of land utilization in the coastal plains of Calabria (South Italy). *GeoJournal* **1986**, *13*, 27–35.

19. Sebastiani, M.; González, S.E.; Castillo, M.M.; Alvizu, P.; Oliveira, M.A.; Pérez, J.; Quilici, A.; Rada, M.; Yáber, M.C.; Lentino, M. Large-scale shrimp farming in coastal wetlands of Venezuela, South America: Causes and consequences of land-use conflicts. *Environ. Manag.* **1994**, *18*, 647–661. [CrossRef]

20. Pérez-Rincón, M.; Vargas-Morales, J.; Crespo-Marín, Z. Trends in social metabolism and environmental conflicts in four Andean countries from 1970 to 2013. *Sustain. Sci.* **2018**, *13*, 635–648. [CrossRef]

21. Carpenter, S.L.; Kennedy, W.J.D. Environmental conflicts in mineral development. *Miner. Environ.* **1980**, *2*, 159–164. [CrossRef]

22. McKenna, J.; Cooper, J.A.G.; O'Hagan, A.M. Coastal erosion management and the European principles of ICZM: Local versus strategic perspectives. *Coast. Conserv.* **2009**, *13*, 165–173. [CrossRef]

23. Tuda, A.O.; Stevens, T.F.; Rodwell, L.D. Resolving coastal conflicts using marine spatial planning. *Environ. Manag.* **2014**, *133*, 59–68. [CrossRef] [PubMed]

24. Zhao, X.; Jia, P. Multi-planning integration advancing coastal zone management: A case study from Hainan Island, China. *Coast. Conserv.* **2018**, 1–7. Available online: https://link.springer.com/article/10.1007/s11852-018-0667-0 (accessed on 30 June 2019). [CrossRef]

25. Rudianto Tantu, A.G. An analysis of coastal land conflict in the North of Jakarta coastal area: (A general algebraic modelling system approach). *Coast. Conserv.* **2014**, *18*, 69–74. [CrossRef]

26. Chuang, Y.H.; Yu, R.F.; Chen, W.Y.; Chen, H.W.; Su, Y.T. Sustainable planning for a coastal wetland system with an integrated ANP and DPSIR model for conflict resolution. *Wetl. Ecol. Manag.* **2018**, *26*, 1015–1036. [CrossRef]

27. Bascetin, A. A decision support system using analytical hierarchy process (AHP) for the optimal environmental reclamation of an open-pit mine. *Environ. Geol.* **2007**, *52*, 663–672. [CrossRef]

28. Sadiq, R.; Tesfamariam, S. Environmental decision-making under uncertainty using intuitionistic fuzzy analytic hierarchy process (IF-AHP). *Stoch. Environ. Res. Risk Assess.* **2009**, *23*, 75–91. [CrossRef]

29. Afshar, A.; Mariño, M.A.; Saadatpour, M.; Afshar, A. Fuzzy TOPSIS Multi-Criteria Decision Analysis applied to Karun Reservoirs System. *Water Resour. Manag.* **2011**, *25*, 545–563. [CrossRef]

30. Pourebrahim, S.; Mokhtar, M.B. Conservation priority assessment of the coastal area in the Kuala Lumpur mega-urban region using extent analysis and TOPSIS. *Environ. Earth Sci.* **2016**, *75*, 348. [CrossRef]

31. Asghari, M.; Nassiri, P.; Monazzam, M.R.; Golbabaei, F.; Arabalibeik, H.; Shamsipour, A.; Allahverdy, A. Weighting criteria and prioritizing of heat stress indices in surface mining using a Delphi Technique and Fuzzy AHP-TOPSIS Method. *Environ. Health Sci. Eng.* **2017**, *15*, 1. [CrossRef]

32. MONRE (Ministry of Natural Resources and Environment). *Basic Contents of Survey on Geology, Environmental Geology and Disasters of Geology in Coastline of Vietnam*; General Department of Geology and Minerals of Vietnam: Hanoi, Vietnam, 2001; 463p. (In Vietnamese)
33. DGM (Department of Geology and Minerals of Vietnam). *Report on Mineral Resources in Coastal Provinces*; Department of Geology and Minerals of Vietnam: Hanoi, Vietnam, 2016; 365p. (In Vietnamese)
34. KAG (Ky Anh Government). *Socio-Economic Development of Ky Anh*; Official Report; KAG: Ky Anh, Vietnam, 2019; 78p. (In Vietnamese)
35. Nguyen, A.T.; Vu, D.A.; Vu, P.V.; Nguyen, T.N.; Pham, T.M.; Nguyen, H.T.T.; Le, H.T.; Nguyen, T.V.; Hoang, L.K.; Duc Vu, T.; et al. Human ecological effects of tropical storms in the coastal area of Ky Anh (Ha Tinh, Vietnam). *Environ. Dev. Sustain.* **2017**, *19*, 745–767. [CrossRef]
36. PMV (Prime Minister of Vietnam). *"Decision 1880/Q-TTg on the Compensation to the Provinces of Ha Tinh, Quang Binh, Quang Tri, and Thue Thien-Hue Following the Marine Environmental Incident"*; PMV: Ha Noi, Vietnam, 2016. (In Vietnamese)
37. PMV (Prime Minister of Vietnam). *"Decision 309/Q-TTg on the revision of Decision 1880/Q-TTg on September 29 2016 Regarding the Compensation for the Provinces of Ha Tinh, Quang Binh, Quang Tri, and Thue Thien-Hue following the Marine Environmental Incident"*; PMV: Ha Noi, Vietnam, 2017. (In Vietnamese)
38. Saaty, R.W. The analytic hierarchy process—What it is and how it is used. *Math. Model.* **1987**, *9*, 161–176. [CrossRef]
39. Vargas, L.G. An overview of the analytical hierarchy process and its applications. *Eur. J. Oper. Res.* **1990**, *48*, 2–8. [CrossRef]
40. Carnero, M.C. Multicriteria model for maintenance benchmarking. *Manuf. Syst.* **2014**, *33*, 303–321. [CrossRef]
41. Chang, D.Y. Applications of the extent analysis method on fuzzy AHP. *Eur. J. Oper. Res.* **1996**, *95*, 649–655. [CrossRef]
42. Pourjavad, E.; Mayorga, R.V. A hybrid approach integrating AHP and TOPSIS for sustainable end-of-life vehicle strategy evaluation under fuzzy environment. *Wseas Trans. Circuits Syst.* **2016**, *15*, 216–223.
43. Hwang, C.L.; Yoon, K. *Multiple Attribute Decision Making: Methods and Applications*; Springer: Berlin/Heidelberg, Germany, 1981; 225p, ISBN 978-3-642-48318-9.
44. Yasmi, Y.; Schanz, H.; Salim, A. Manifestation of conflict escalation in natural resource management. *Environ. Sci. Policy* **2006**, *9*, 538–546. [CrossRef]
45. Taylan, O.; Bafail, A.O.; Abdulaal, R.M.S.; Kabli, M.R. Construction projects selection and risk assessment by fuzzy AHP and fuzzy TOPSIS methodologies. *Appl. Soft Comput.* **2014**, *17*, 105–116. [CrossRef]
46. Sirisawat, P.; Kiatcharoenpol, T. Fuzzy AHP-TOPSIS approaches to prioritizing solutions for reverse logistics barriers. *Comput. Ind. Eng.* **2018**, *117*, 303–318. [CrossRef]
47. Tyagi, M.; Kumar, P.; Kumar, D. A hybrid approach using AHP-TOPSIS for analyzing e-SCM performance. *Procedia Eng.* **2014**, *97*, 2195–2203. [CrossRef]
48. Sun, L.; Zhu, D.; Chan, E.H.W. Public participation impact on environment NIMBY conflict and environmental conflict management: Comparative analysis in Shanghai and Hong Kong. *Land Use Policy* **2016**, *58*, 208–217. [CrossRef]
49. Rincón-Ruiz, A.; Rojas-Padilla, J.; Agudelo-Rico, C.; Perez-Rincon, M.; Vieira-Samper, S.; Rubiano-Paez, J. Ecosystem services as an inclusive social metaphor for the analysis and management of environmental conflicts in Colombia. *Ecosyst. Serv.* **2019**, *37*, 100924. [CrossRef]
50. Koornneef, J.; Faaij, A.; Turkenburg, W. The screening and scoping of environmental impact assessment and strategic environmental assessment of carbon capture and storage in the Netherlands. *Environ. Impact Assess. Rev.* **2008**, *28*, 392–414. [CrossRef]
51. Nguyen, A.T.; Vu, A.D.; Dang, G.T.H.; Hoang, A.H.; Hens, L. How do local communities adapt to climate changes along heavily damaged coasts? A Stakeholder Delphi study in Ky Anh (Central Vietnam). *Environ. Dev. Sustain.* **2018**, *20*, 749–767. [CrossRef]

 *applied sciences*

*Article*

# Microbial Induced Carbonate Precipitation Using a Native Inland Bacterium for Beach Sand Stabilization in Nearshore Areas

Pahala Ge Nishadi Nayanthara [1,*], Anjula Buddhika Nayomi Dassanayake [2], Kazunori Nakashima [3] and Satoru Kawasaki [3]

[1] Graduate School of Engineering, Hokkaido University, Kita 13, Nishi 8, Kita-Ku, Sapporo, Hokkaido 060-8628, Japan
[2] Faculty of Engineering, University of Moratuwa, Moratuwa 10400, Sri Lanka
[3] Faculty of Engineering, Hokkaido University, Kita 13, Nishi 8, Kita-Ku, Sapporo, Hokkaido 060-8628, Japan
[*] Correspondence: nishadi.nayanthara@gmail.com; Tel.: +81-80-2123-1764

Received: 4 July 2019; Accepted: 2 August 2019; Published: 6 August 2019

**Featured Application: Coastal erosion is a natural process which poses serious problems to coastal communities worldwide. Hard engineering solutions are the most common defense mechanisms that have been used for a long time; although attention has now been inevitably given towards more ecologically and economically sustainable soft engineering approaches. Through this study, it is intended that a bioengineered solution be proposed against coastal erosion using bacterial biomineralization. However, the results presented in this paper are only from a bottom-line study and much more realistic investigations, including scaling up, should be carried out in the future.**

**Abstract:** Microbial Induced Carbonate Precipitation (MICP) via urea hydrolysis is an emerging sustainable technology that provides solutions for numerous environmental and engineering problems in a vast range of disciplines. Attention has now been given to the implementation of this technique to reinforce loose sand bodies in-situ in nearshore areas and improve their resistance against erosion from wave action without interfering with its hydraulics. A current study has focused on isolating a local ureolytic bacterium and assessed its feasibility for MICP as a preliminary step towards stabilizing loose beach sand in Sri Lanka. The results indicated that a strain belonging to *Sporosarcina* sp. isolated from inland soil demonstrated a satisfactory level of enzymatic activity at 25 °C and moderately alkaline conditions, making it a suitable candidate for target application. Elementary scale sand solidification test results showed that treated sand achieved an approximate strength of 15 MPa as determined by needle penetration device after a period of 14 days under optimum conditions. Further, Scanning Electron Microscopy (SEM) imagery revealed that variables such as grain size distribution, bacteria population, reactant concentrations and presence of other cations like $Mg^{2+}$ has serious implications on the size and morphology of precipitated crystals and thus the homogeneity of the strength improvement.

**Keywords:** beach sand stabilization; coastal erosion; local ureolytic bacterium; microbial induced carbonate precipitation; crystal morphology; nearshore areas

## 1. Introduction

Microbial Induced Carbonate Precipitation (MICP) through a ureolytic pathway is nowadays a subject of intense research interest in the field of biogeotechnology as it provides solutions to a wider range of engineering applications. Several studies have demonstrated that MICP can be employed to improve the mechanical properties of weak porous materials [1] and cater to a number of technological applications including ground reinforcement, slope stabilization, liquefaction prevention, erosion prevention of dams and levees, toxic metals removal from contaminated wastewaters/waters, immobilization of hazardous contaminants in soil, and $CO_2$ sequestration, among others [2–16]. Recently, this mechanism has been suggested as a novel approach towards erosion mitigation of sandy soils in nearshore regions as it is a minimal disturbance to natural coastal environment.

Loose beach sands generally show high erodibility under strong wave action, rising sea levels, as well as under strong wind currents closer to the surface. Conventional engineering solutions such as construction of sea walls, dykes, embankments, groins, and breakwaters have succeeded in combatting the beach sand erosion problem over decades [17,18]. However, they are associated with continual and costly maintenance work, large visual impact lowering the recreational values of beaches and downdrift erosion in the presence of alongshore currents [18]. Chemical grouting is another commonly applied technique in civil engineering to stabilize weaker sands/soils. However, most of the chemical grouts are toxic to humans and require comparatively high cost [1]. Greener solutions such as vegetation belts, sand supply, and artificial dunes are not primary solutions but mere additional measures to protect beaches [17].

The myriad of issues associated with conventional engineering solutions have necessitated a search for alternative measures and, markedly, MICP has gained a lot of attention due to its economical and ecological sustainability. Albeit in limited amounts, previous studies have already shown possible avenues for loose beach sand stabilization for coastal zone management through MICP-based biogeotechnical approaches [19–23]. MICP is based on a set of complex biochemical reactions which ultimately produce calcium carbonate ($CaCO_3$) crystals that physically bond the loose sand grains together thereby improving mechanical and geotechnical engineering properties of treated sand [24]. In MICP, microbial enzyme urease (urea amidohydrolase; EC 3.5.1.5) catalyzes hydrolysis of urea ($CO(NH_2)_2$) to ammonia and carbonic acid ($H_2CO_3$) (Equations (1) and (2)). The resulting ammonia and carbonic acid equilibrate to form $HCO_3^-$ (Equation (3)) and $NH_4^+$ (Equation (4)) with equilibrium constants of $pK_1$ 6.3 and $pK_a$ 9.3, respectively [25]. This phenomenon creates a high carbonate alkalinity in the micro environment around the cell [26] which favors the formation of carbonate ions (Equation (5)). In the presence or supply of a calcium source, $Ca^{2+}$ ions will be attracted towards negatively charged bacterial cell walls, resulting a localized state of super saturation with respect to calcium carbonate. Subsequently, calcium carbonate will be precipitated on cell surfaces which also act as nucleation sites (Equation (6)) once a certain level of supersaturation is achieved [13].

$$CO(NH_2)_2 + H_2O \xrightarrow{Urease} NH_3 + NH_2COOH \tag{1}$$

$$NH_2COOH + H_2O \longrightarrow H_2CO_3 + NH_3 \tag{2}$$

$$H_2CO_3 \longrightarrow H^+ + HCO_3^- \tag{3}$$

$$NH_3 + H_2O \longrightarrow NH_4^+ + OH^- \tag{4}$$

$$HCO_3^- \longrightarrow H^+ + CO_3^{2-} \tag{5}$$

$$CO_3^{2-} + Ca^{2+} \xrightarrow{Bacterial\ cells} CaCO_3 \downarrow \tag{6}$$

The enhancement in the mechanical behavior of MICP treated sands is accomplished by a few physical mechanisms based on how calcium carbonate is precipitated with respect to the particles. Calcium carbonate crystals can grow on grain surfaces individually coating them (grain coating), or may precipitate at or near contact points of particles (contact cementing), or grow into the void spaces from particle surfaces, subsequently creating calcium carbonate bridges among sand grains (matrix-supported) [27]. Further, calcium carbonate cements can grow, filling the pore spaces without bridging the sand grains thus lowering the hydraulic conductivity of the treated samples. However, the overall improvement in the mechanical properties of loose sand following MICP treatment is due to a combination of all four aforementioned cementation mechanisms (Figure 1).

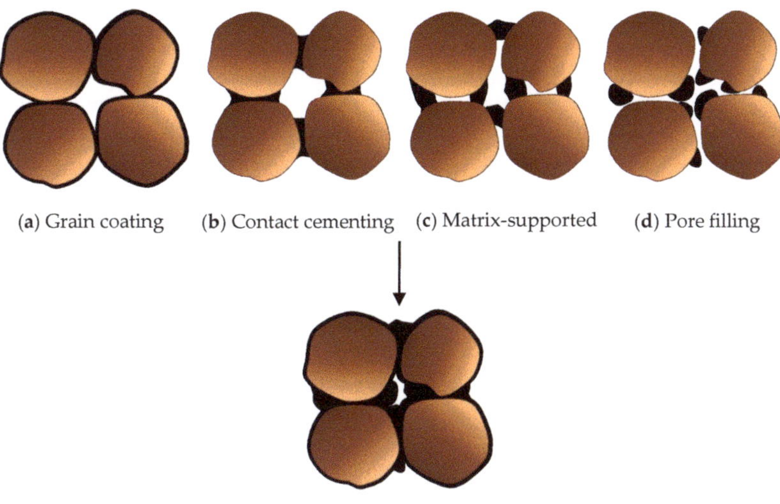


**Figure 1.** Cementation mechanism of microbial induced carbonate precipitation (**a**) grain coating; (**b**) contact cementing; (**c**) matrix-supported; (**d**) pore-filling.

Moreover, MICP can be harnessed for in-situ sand/soil applications in two different approaches: biostimulation and bioaugmentation. Biostimulation is the process of modifying prevailing environmental conditions (substrates, nutrients) to encourage calcium carbonate precipitation through enrichment of natural indigenous urea hydrolyzing bacteria [5,28]. Although stimulated ureolytic species are more resilient for in-situ applications, their urea hydrolysis rates seem to be lower than the specialized laboratory cultivated strains [5], and level of MICP will be limited to the initial bacterial cell concentration in the soil [29]. In contrast, bioaugmentation is implemented by supplementing the soil of concern with bacterial strains having specific metabolic capabilities along with nutrient and precipitation media [28,30]. Generally, in-situ bioaugmentation with exogeneous strains is challenging as injected foreign cultures may face competition for survival and proliferation among native communities along with poor compatibility in new environments [25,30,31]. Therefore, a more successful approach would be to augment the soil to be treated with native ureolytic bacteria that has been isolated and enriched in laboratory culture media in advance.

The main objective of the present study is to evaluate the effectiveness of MICP treatment via bioaugmentation to stabilize loose beach sand in Sri Lanka, an island highly vulnerable to coastal erosion. As an initial step towards achieving this goal, native ureolytic strains were isolated from Sri Lankan sand/soil and assessed for their potential for bioaugmentation. Upon receiving more successful results for MICP treatment using an inland ureolytic bacterium than that from local marine bacteria for the desired engineering application, further studies were carried out to (i) optimize the growth and enzymatic activity of the isolated strain, (ii) address mechanical behavior and microstructure of the MICP-treated sand and (iii) explore the versatility of the isolated inland strain to perform biomediated calcium carbonate precipitation in nearshore environments as an alternative technique for sand stabilization.

## 2. Materials and Methods

### 2.1. Isolation and Identification of a Suitable Urease Active Bacterium

#### 2.1.1. Soil Sampling and Isolation of Ureolytic Bacteria

The coastal hydrodynamics along the southern, south western, and western coastal belt of Sri Lanka are highly affected by the strong waves generated by the south western monsoon and are the most susceptible to erosion of sandy slopes. Therefore, the study area for the current study was demarcated from Tangalle in southern province to Uswetakeiyawa, Colombo, in western province. Beach sand samples, each approximately weighing 50 g, were randomly collected from 11 sampling locations (M1–M11) along the shoreline and soil samples from 2 inland localities (L1 and L2) into sterile tubes (see Supplementary Materials). All the samples were exported from Sri Lanka to Japan following the Japanese Minister Permission System and stored at 4 °C. Each soil sample of 5 g was then serially diluted ($10^{-1}$–$10^{-6}$ times) under sterile conditions. 50 μL of beach sand and inland soil suspensions thus prepared were then plated on Zobell 2216E agar medium (5.0 g/L hipolypeptone, 1.0 g/L yeast extract, 0.1 g/L $FePO_4$, 30.0 g/L agar in artificial seawater, pH 7.6–7.8) and $NH_4$-YE agar medium (15.75 g/L tris buffer, 10.0 g/L $(NH_4)_2SO_4$, 20.0 g/L yeast extract, 20.0 g/L agar in distilled water), respectively. After incubating at 30 °C for 2 days, colonies were identified from a plate with 30–200 colonies. Different types of colonies were visually identified and separately cultivated on the plates prepared in the same manner described above.

A simple urease activity test was then conducted for qualitative assessment of urease producing bacteria. Each colony was mixed with 20 mL of cresol red solution containing urea (0.4 g/L cresol red and 25 g/L $CO(NH_2)_2$ in distilled water) and incubated at 45 °C for two hours. Colonies that changed the initial yellow color to pink (Cresol red changes from yellow to pink when pH changes to 7.2–8.8, which is accomplished during urea hydrolysis) were identified as urease active bacteria.

#### 2.1.2. Identification of Bacteria by 16S rRNA Sequencing

16S rRNA gene amplification and sequencing was carried out for selected ureolytic isolates. The analysis of the DNA sequences was performed by using the DB-BA 12.0 (Techno Suruga Lab Co., Ltd., Shizuoka, Japan) and International Nucleotide Sequence Database (DDBJ/ENA (EMBL)/GenBank) by Techno Suruga Laboratory, Shizuoka, Japan.

#### 2.1.3. Cultivation of Bacteria and Assessing the Potential for Microbial Induced Carbonate Precipitation

Zobell 2216E for marine bacteria and $NH_4$-YE medium for land bacteria were used for culturing under sterile aerobic conditions. The cells were precultured separately in 5 mL of each media at 30 °C and 160 rpm for 24 h. One mL of the preculture was inoculated with 100 mL of the fresh medium and incubated under the same conditions. Microbial cell growths of the isolates were determined in terms of optical density at a wave length of 600 nm ($OD_{600}$) using a UV–vis spectrophotometer (V-730, JASCO Corporation, Tokyo, Japan) from representative specimens sampled at regular intervals of 24 h.

Microbial induced calcium carbonate precipitation tests were conducted in 10 mL tubes to evaluate feasibility of using isolated bacterial strains for biomediated soil improvement. Calcium chloride ($CaCl_2$) reagent was chosen as the calcium source and 1 mL from equimolar concentrations (0.5 mol/L (M)) of $CaCl_2$ and urea solutions were used. One mL from the bacteria culture was added to the tube and total volume was adjusted to 10 mL using distilled water. Samples were then kept in a shaking incubator at 30 °C and 160 rpm for 48 h. Finally, the reaction mixtures were centrifuged and precipitates were collected to gravimetrically compare the amount of precipitate formed by each isolate. Based on the results, a single isolate was picked as the most suitable for further analysis.

### 2.1.4. Microbial Cell Growth and Urease Activity Measurement

Urease activity of the selected isolate was evaluated by cells suspended in a solution containing urea in terms of the amount of enzyme required to hydrolyze one micromole of urea per minute (U) per milliliter following the method explained by Gowthaman et al. [3]. The test was repeated at 24 h intervals to determine the temporal stability of the urease enzyme for bacteria cultured at different temperatures ranging from 20–50 °C.

In order to monitor the pH dependency of the enzymatic activity, the urease activity test was performed for 1 mL samples collected from a bacterial cell culture incubated at the optimum temperature (decided from the above step) and subsequently suspending them in substrate solutions prepared with different pH values (ranging from pH 5–9).

### 2.2. Small Scale Sand Solidification Test

Centimeter-scale one-dimensional column specimen tests under different experimental conditions were performed as a preliminary step towards elucidating the potential for strengthening and stabilizing loose beach sand by MICP.

Two types of natural beach sands with varying particle size gradations (see Supplementary Materials) collected from Sri Lanka were used in the current study; a well graded sand and a uniformly graded sand (herein after referred to as sand 1 and sand 2, respectively). The particle size distribution curves were obtained by sieve analysis using BS410 standard sieves. The characteristics and chemical composition of the sands (as determined from powdered samples using Energy Dispersive X-ray Fluorescence spectrometer (JSX-31000R II JEOL, Ltd., Tokyo, Japan) at Material Analysis and Structural Analysis Open Unit of Hokkaido University, Japan) are tabulated in Table 1.

**Table 1.** Characteristics of natural beach sand used in the current study and their elemental composition.

| Sand Type | $\varrho$ | $D_{50}$ | $C_u$ | $C_c$ | Elemental Composition (%) | | | | | | | | | |
|---|---|---|---|---|---|---|---|---|---|---|---|---|---|---|
| | | | | | $Al_2O_3$ | $SiO_2$ | CaO | $TiO_2$ | $Cr_2O_3$ | $Fe_2O_3$ | NiO | CuO | SrO | PbO |
| Sand 1 | 2.81 | 0.37 | 3.58 | 1.12 | 3.64 | 75.39 | 4.25 | 8.39 | 0.04 | 8.13 | 0.08 | 0.07 | - | - |
| Sand 2 | 2.68 | 0.61 | 1.51 | 0.93 | - | 24.14 | 71.79 | 0.27 | - | 2.98 | 0.04 | - | 0.61 | 0.06 |

$\varrho$—specific gravity; $D_{50}$—diameter at which 50% of the sample's mass is comprised of smaller particles than that value; $C_u$—coefficient of uniformity; $C_c$—coefficient of curvature.

The optical density ($OD_{600}$) of the harvested culture varied between 2.0 and 2.5 and all the solidification tests were conducted at an optimum temperature of 25 °C for a treatment period of 14 days. 0.5 M and 1.0 M $Ca^{2+}$ cementation solutions prepared with distilled water were used to examine the effect of $Ca^{2+}$ ion concentration on solidification of sand. The 0.5 M cementation solution composed of 30.0 g/L of urea ($CO(NH_2)_2$), 55.0 g/L of $CaCl_2$, and 3.0 g/L of nutrient broth.

A standard size 35 mL syringe (2.5 cm diameter) was used for the sand solidification tests. In each case, sand was oven dried at 105 °C for 48 h and a predetermined weight of sand was compacted in 3 layers by applying 20 hammer blows on each layer. The bottom face of each column was blocked by a filter paper.

After considering the studies pertaining to determine the most favorable and proper treatment method for MICP applications, a method involving cycles of batch treatment of bacterial culture and cementation solution was adopted for this study. Firstly, a 12 mL of bacterial culture was injected from the top of the syringe, excess solutions were drained out at a controlled rate and allowed approximately 2 h for bacteria fixation. Secondly, this was followed by 12 mL of cementation solution, supplying the necessary nutrient and precipitation components.

Moreover, in terms of saturation level, the entire length of the sand column was retained under fully saturated condition as an initial test condition. Four test cases were carried out, varying the reactant concentration and frequency of bacteria injection. For cases 1 and 3, the bacterial culture solution was injected only in the beginning (day 0) whereas for cases 2 and 4, the solution was injected twice, on day 0 and day 7. Cementation solution was injected every 24 h for the total treatment period. However, 0.5 M cementation solution was injected for test cases 1 and 2 while 1.0 M solution was supplied to test cases 3 and 4.

*2.3. Effect of $Mg^{2+}$ on Microbial Induced Carbonate Precipitation*

The majority of the MICP-based studies carried out so far has assessed the change in soil properties with calcium carbonate precipitation. However, the real field conditions are such that it is likely to contain other impurities and cations that may interact with the precipitation process. For the implementation of MICP for the desired application in nearshore environments, attention has to be paid to the fact that target treatment environments will consist of a considerable amount of magnesium ions. Therefore, a set of experiments were conducted incorporating both $Mg^{2+}$ and $Ca^{2+}$ ions with various $Mg^{2+}/Ca^{2+}$ molar ratios from 0 to 1.

Anhydrous magnesium chloride was used as the $Mg^{2+}$ source. Microbial induced carbonate precipitation tests were carried out as described above, but according to the test program in Table 2. Sand solidification tests were done using "sand 2" under fully saturated conditions for the same $Mg^{2+}/Ca^{2+}$ molar ratios employed in precipitation tests.

**Table 2.** Microbial induced carbonate precipitation and sand solidification test conditions with magnesium addition.

| Test Case | $Mg^{2+}$ (M) | $Ca^{2+}$ (M) | $Mg^{2+}/Ca^{2+}$ Ratio | Sum of $Mg^{2+}$ and $Ca^{2+}$ (M) | Urea (M) |
|---|---|---|---|---|---|
| 1 | 0.00 | 0.50 | 0.00 | 0.50 | 0.50 |
| 2 | 0.10 | 0.40 | 0.25 | 0.50 | 0.50 |
| 3 | 0.15 | 0.35 | 0.43 | 0.50 | 0.50 |
| 4 | 0.25 | 0.25 | 1.00 | 0.50 | 0.50 |

*2.4. Monitoring Methods*

2.4.1. Needle Penetration Device

The local cementation strength of the specimens along the height of the column was examined in terms of estimated unconfined compressive strength (UCS) using a soft rock penetrometer (SH-70, Maruto Testing Machine Company, Tokyo, Japan). The needle of the device was penetrated into the sand specimens at three locations (1, 3, and 5 cm from the bottom of the column) and the penetration resistance (N) and penetration distance (mm) were simultaneously measured. From the results, a "penetration gradient" (N/mm) was determined, which was ultimately used to estimate UCS using Equation (7) where $x$ is the penetration gradient and $q_u$ is the UCS.

$$\log(q_u) = 0.978 \log(x) + 2.621 \tag{7}$$

The regression relation in Equation (7) has been developed by analyzing an adequate number of natural soft rock samples and cement based improved soils and confirmed by a correlation coefficient of 0.914.

2.4.2. Scanning Electron Microscopy (SEM) Analyses

Fractions of cemented sand collected from different parts of the oven dried column were observed under a Scanning Electron Microscope (SuperScan SS-550, Shimadzu Corporation, Kyoto, Japan) to study the soil matrix biocemented with calcium carbonate.

### 2.4.3. X-ray Diffraction (XRD) Analyses

X-ray diffractometer patterns of the powdered precipitates from Section 2.3 were analyzed using an X-ray generator (MiniFlex™, Rigaku Co., Ltd., Tokyo, Japan). The analysis was carried out at room temperature using the XRD machine with CuKa radiation at a rate of 6.5° 2θ/min and a step size of 0.02° 2θ ranging from 10 to 80° 2θ. Qualitative mineralogy of the samples was determined with the standard interpretation procedures of XRD using Match! software for phase identification from powder diffraction.

### 2.4.4. X-ray Fluorescence (XRF) Analyses

The quantitative elemental analysis of the powdered sand samples was determined using an Energy Dispersive X-ray Fluorescence spectrometer (JSX-31000R II JEOL, Ltd., Tokyo, Japan) at the Material Analysis and Structural Analysis Open Unit of Hokkaido University, Japan. The instrument consisted of a 5–50 kV, 1 mA, 50 W X-ray generator. Samples were observed by a colorful CCD camera.

## 3. Results and Discussion

### 3.1. Isolation and Identification of a Suitable Urease Active Bacterium

Although a total of more than 100 bacterial strains were isolated from different soil samples, urease producing bacteria were encountered in only 6 locations. Apparently, the proportion of ureolytic bacteria in each case was smaller than that of non-ureolytic bacteria except for location L2 (Figure 2). As per these results, it is evident that ureolytic bacteria are not ubiquitous in the dynamic coastal regions as much as in the inland environments. Based on the simple urease activity test and the corresponding increase in pH in reaction mixture during first 24 h (data not included), 6 isolates were considered for further analysis. The phylogenetic groupings of the chosen isolates are shown in Table 3.

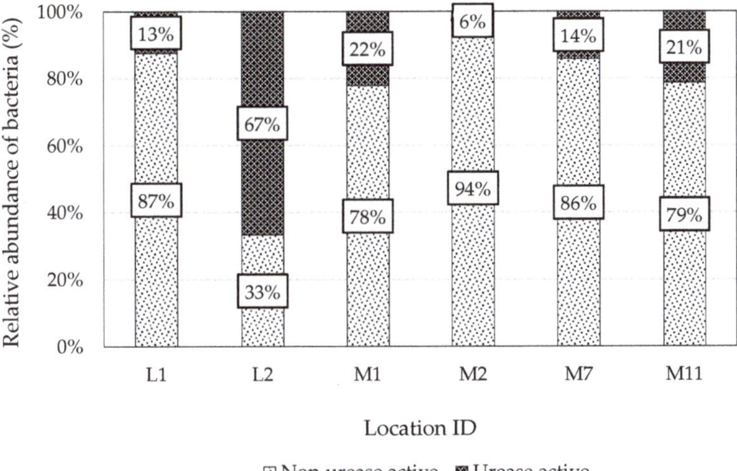


**Figure 2.** Relative abundance of ureolytic bacteria in different sampling locations.

**Table 3.** Phylogenetic grouping of the isolated bacterial strains.

| Sample ID | Closest Match to Phylogenetic Grouping | Genbank Accession Number | Similarity |
|---|---|---|---|
| M1-1-1 | *Halomonas* sp. | AJ306891 | 99.8% |
| M2-2-1 | *Halomonas* sp. | AJ306888 | 99.6% |
| M11-1-5 | *Sulfitobacter* sp. | AY180103 | 98.0% |
| M11-1-7 | *Oceanobacillus* sp. | HQ595230 | 99.9% |
| L2-2-2 | *Sporosarcina* sp. | HQ676600 | 99.7% |
| L1-1-5 | *Bacillus* sp. | NR_146034 | 98.7% |

In terms of $OD_{600}$, all the isolates showed a stable growth at 30 °C. However, the strain belonging to *Sporosarcina* sp. preferably witnessed a more stable growth over the entire test period (see Supplementary Materials).

The test tube precipitation test results are a good relative indication of the potential of each strain for sand solidification by MICP [32] as the strength improvement is proportional to the amount of bio-precipitate formed. The results indicated that the strain belonging to *Sporosarcina* sp. shows the highest performance for microbial induced carbonate precipitation and the weight of precipitate is approximately twice as that from others (Figure 3). The strains belonging to *Sporosarcina* sp. strain are gram stain positive, rod shaped, motile, aerobic nonpathogenic bacterium which produces active intracellular urease in large quantities [13,33–37].

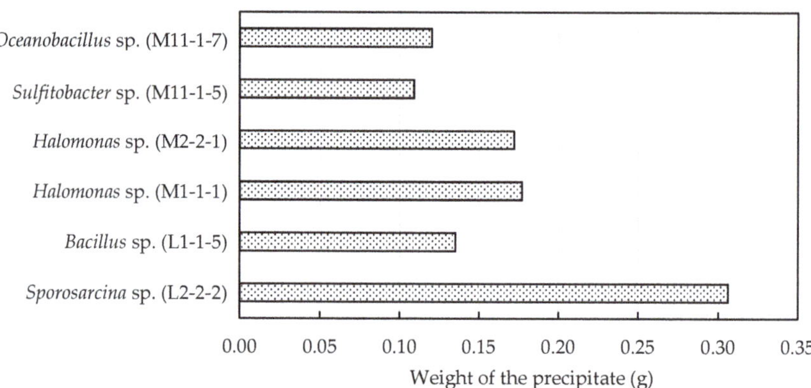


**Figure 3.** Microbial induced carbonate precipitation capacity of isolated strains.

On the other hand, strains belonging to *Sporosarcina* sp., specially *Sporosarcina pasteurii*, has been extensively studied for a vast variety of MICP applications and has shown successful results [13,15,16,33,35,38–42]. More interestingly, this specific strain has the ability to tolerate extreme conditions, for example pH levels as high as 9 [43], making it a good candidate for the target application. Considering all these facts, the *Sporosarcina* sp. strain was identified as the most suitable bacterium, and remaining laboratory investigations were carried out employing this strain.

*3.2. Optimizing the Growth and Enzymatic Activity of the Isolated Sporosarcina sp. Strain*

3.2.1. Effect of Temperature

In order to ascertain the maximum benefits from the MICP process, it is important to determine the temperature at which microbial growth and urease activity is optimum. As per the findings of this study, the microbial growth of the isolate was more stable and comparatively higher at a temperature range of 20–25 °C whereas the growth became almost zero when the temperature was raised up to 50 °C (Figure 4). Ng et al. [12] states that the microbial growth is generally less sensitive to temperature changes over the range of 20–30 °C, which was also true in this case. In addition, the strain exhibits the typical growth curve of a bacteria population over time where cell growth increases exponentially up to a maximum followed by a stationary phase and then a gradual fall (death phase) due to independent or combined effect of nutrient exhaustion, waste accumulation and/or oxygen depletion, and subsequent cell lysis [44].

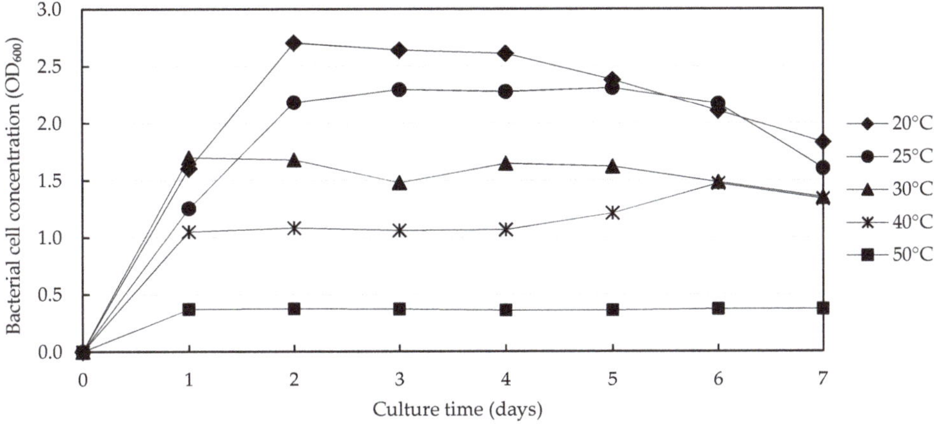


**Figure 4.** Variation of bacterial cell concentration of *Sporosarcina* sp. at different temperatures.

Catalysis of urea hydrolysis is a temperature dependent enzymatic reaction [43] and is closely associated with the bacterial cell population [12]. The effect of temperature on urease activity of the isolate is presented in Figure 5 indicating that enzymatic performance is maximum and stable at 25 °C. The activity is degraded almost to zero at 40 °C and 50 °C. This is in accordance with the findings of Dhami et al. [45] where microbial enzyme activity of *Bacillus megaterium* decreased by a fraction of 47% when temperature was raised from 35 to 55 °C. In contrast to that, Sahrawat [46] in a very early study revealed that optimum urease activity lies at temperatures close to 60 °C. This was later attested to by Liang et al. [47] and few others for strains *Streptococcus salivarius* [48] and *Parahodobacter* sp. [49]. In terms of temporal stability, although the selected bacterium maintains a relatively satisfactory enzyme stability over a period of 7 days, the activity seems to drop over time as cell lysis and denaturation of the loose enzyme take place [50]. However, considering the above results as well as the tropical climate in Sri Lanka, the remaining investigations were carried out at a temperature of 25 °C.

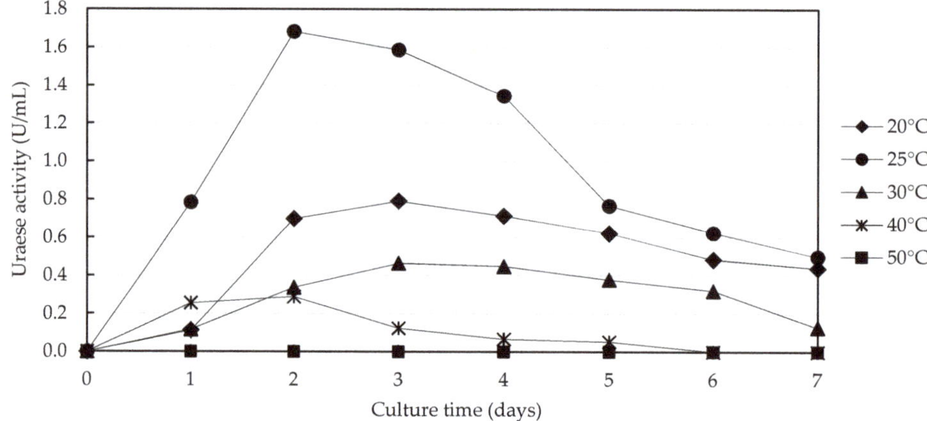

Figure 5. Variation of urease activity of *Sporosarcina* sp. at different temperatures.

### 3.2.2. Effect of pH

pH has a significant influence on urease activity and, hence, the success of MICP. Under highly acidic or alkaline conditions, the urease enzyme is generally irreversibly denatured [49,51]. Figure 6 depicts the degree of change in urease activity at different pH levels. It is evident from the results that, urease activity increases rapidly from pH 6.0 to 9.0 and the enzyme is comparatively more active in the pH range 7.5–9.0. Mobley et al. [51] reported that majority of the microbial ureases, except for a few acid ureases, are optimum under neutral pH conditions. However, the findings of Stocks-Fischer et al. [13] are on the same trend as this study: They reported that urease activity of *Sporosarcina pasteurii*, drastically increased from pH 6.0 to 8.0 and gradually reduced thereafter. Nevertheless, the activity at pH 9.0 for that specific strain was still satisfactory for MICP applications. Ciurli et al. [52] are other researchers who have shown that optimum pH for urease activity of *S. pasteurii* is 8.0. Moreover, few other microbial strains have showed their optimum performance at pH 8.0, as reported in literature [49,53]. From the aforementioned results, it is clear that the isolate used in this study is moderately alkalo-tolerant and can be successfully employed for the target application in a coastal environment.

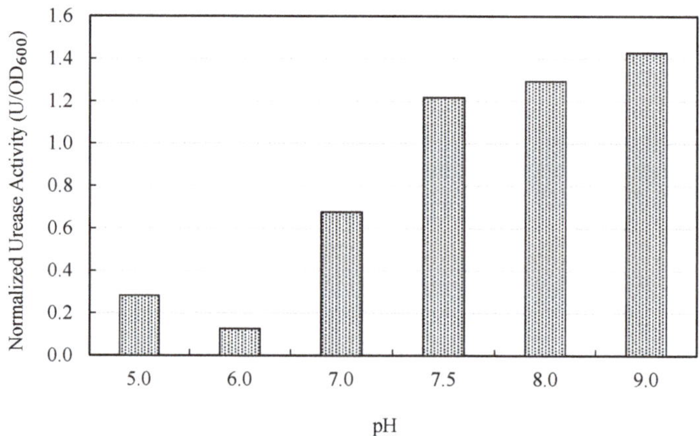

Figure 6. Variation of urease activity of *Sporosarcina* sp. at different pH conditions.

### 3.3. Potential for Beach Sand Solidification by MICP Treatment

### 3.3.1. Effect of Bacteria Population and Reactant Concentration

The reactant and the bacteria cell concentration are two primary factors that governs the degree and the homogeneity of the strength increment over the treatment depth. The local strength of sand specimens underwent different treatment conditions with respect to the aforementioned two parameters, are illustrated in Figure 7. However, the cementation reagent contained equimolar concentrations of urea and $CaCl_2$ as in Ng et al. [54]. All the samples were strongly cemented and the estimated average (as well as individual) UCS values from two independent tests for all the cases were above 3 MPa. When the two samples injected with 0.5 M cementation solution were considered, it was clear that the specimen with bacteria injected twice showed a significant strength increment than that of the specimen injected with bacteria only in the beginning. According to previous records, it is deduced that higher concentrations of bacteria populations enhance the amount of $CaCO_3$ precipitate and, thus, the results of MICP treatment [41]. Generally, bacteria play two key roles in the biocementation process; (i) catalysis of urea hydrolysis and (ii) provision of nucleation sites [55]. The initial investigations on urease activity of the isolated strain showed that enzymatic activity diminishes over time. Besides, there are several other factors that may gradually inhibit the bacteria's affinity to perform their intended role as a catalyst for urea hydrolysis over the entire treatment time period. Possible reasons are hydraulic constraints such as cell encapsulation from the resulting precipitate or being confined inside pores, restricted movement of nutrients and precipitation components as pore spaces get narrowed after initial precipitation, space constraints in the case of saturated conditions, accumulation of metabolic wastes and also flushing out of bacteria [50,56]. Further, bacterial cell walls act as nucleation sites for $CaCO_3$ precipitation in biochemical processes [13,57] and some of the above mentioned factors may lower the availability of cells that can serve this purpose. Therefore, it is concluded here that the rate of urea hydrolysis and subsequent biocementation was retarded over time in the case where bacteria was injected only in the beginning, whereas the urease activity was regained and number of nucleation sites were improved with the reinjection in the case where culture was injected twice, leading to a better performance. This phenomenon was further confirmed by the measurement of $Ca^{2+}$ concentration and the pH of the samples collected from the solution draining out from the outlet (Figure 8). Better performance of microbial induced carbonate precipitation was characterized by lower $Ca^{2+}$ concentration and higher pH values as soon as the bacterial culture was injected.

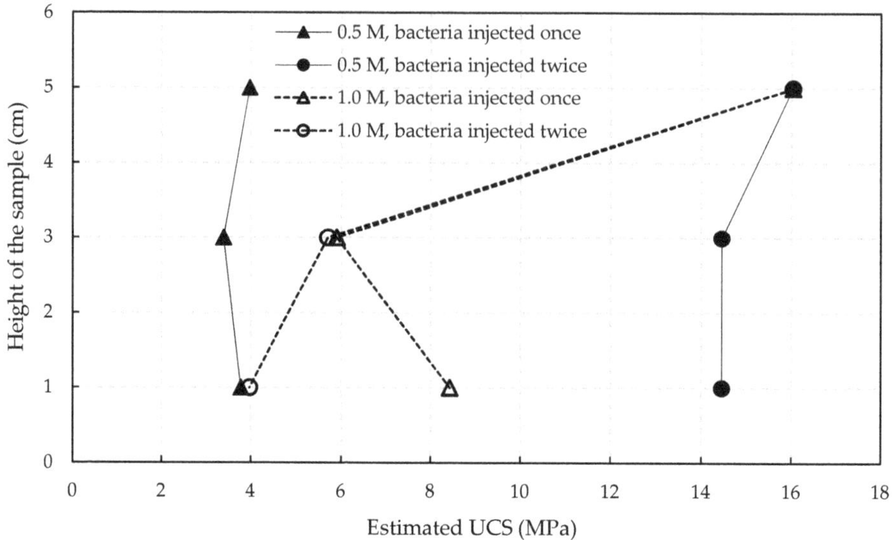

**Figure 7.** Variation of local strength of microbial induced carbonate precipitation (MICP) treated sand under different bacteria and reactant concentrations.

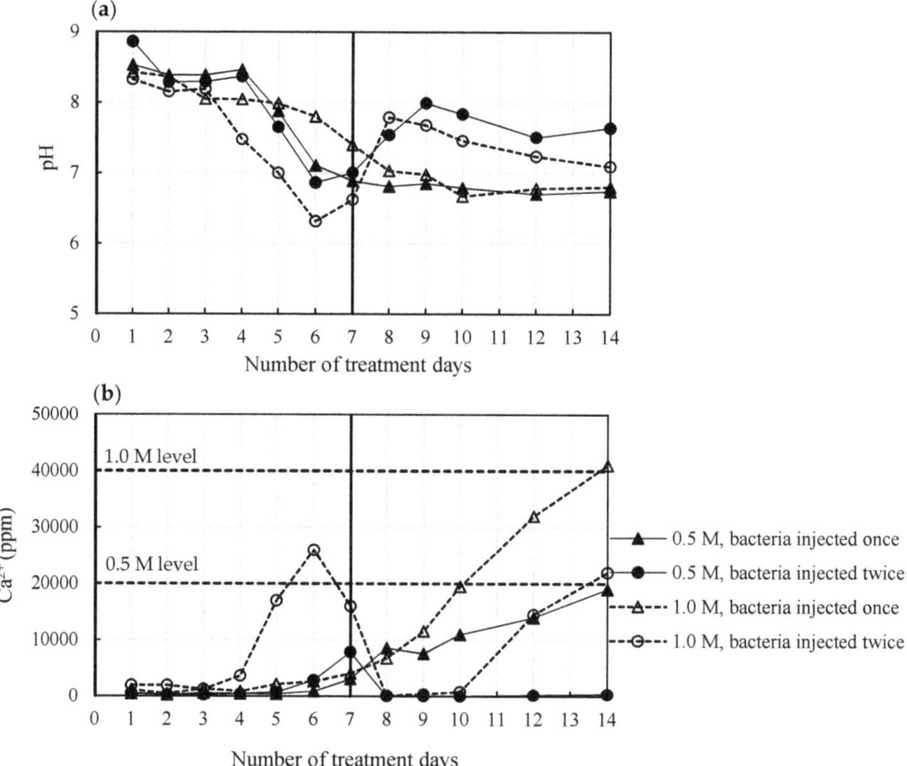

**Figure 8.** (**a**) pH; (**b**) $Ca^{2+}$ measurements of the solution collected from the bottom of the sand.

In contrary to that, samples treated with 1.0 M cementation solution showed a noteworthy heterogeneity in strength distribution over the length of the column as well as among the two duplicated samples. The top portion of the specimens gained strengths as similar to or remarkably higher than their 0.5 M equivalent. However, middle and bottom portions did not harden as much as the top, and all specimens supplied with 1.0 M solution had a greater tendency to clog. The strength variation vertically along the column is suspected to happen due to several reasons. One possibility is filtering and a better accumulation of bacterial cells on top and depletion of reactant concentration as they are been readily consumed for biocementation while flowing from top to bottom [50]. This filtering mechanism is more significant in this case as the virgin beach sand used here consists of a significant percentage of fines (10.5% fines smaller than 125 μm). Additionally, the precipitated calcium carbonate crystals may act as a filter and trap bacterial cells from being washed out to the lower portions [7]. The other pertinent reason is the inhibition of the activity of the aerobic bacterium as the availability of oxygen diminishes from top to bottom [58]. The combination of these phenomena is suspected to impair the efficiency of MICP far away from the injection point.

Further, the variation in strength due to the effects of input bio-chemical concentration was investigated in detail by SEM imaging. Figure 9 highlights the differences in the microfeatures of the crystals precipitated for 0.5 M and 1.0 M cementation solutions. Overall, for all the treatment variations, rhombohedral calcium carbonate crystals were predominantly observed under SEM which is consistent with the previous findings [27,35,41,59].

For the 0.5 M cases, the $CaCO_3$ crystals showed a uniform precipitation pattern where the grains were covered with similar sized crystals which were very well distributed spatially on grain surfaces over the entire length of the sample. The contact areas were also covered uniformly and for the case with two bacteria injections, contact cementing and matrix supporting in the column top was characterized with larger crystals of 20–25 μm in size (Figure 9a). This pattern gave rise to a highly favorable homogeneous strength improvement in the column making it the most suitable treatment condition. But, for the case with single bacterial culture treatment, although uniformity was still maintained, only thin layers of 2–3 μm ($CaCO_3$ envelopes) covering individual sand particles and slightly bridging the adjacent grains were achieved which did not provide a satisfactory strength improvement.

In the case of 1.0 M treatments, although grain surfaces were predominantly covered with uniformly sized crystals, the precipitation had become more random as more calcite formed. Larger crystals of approximately 30 μm in size were randomly distributed, filling the pore spaces in the top portion and clogging near the injection point (Figure 9d). Ng et al. [54] stated that flow would be obstructed even without any specific binding of soil particles as long as precipitated crystals are larger than the pore throat. Moreover, only thin layers of differently sized $CaCO_3$ crystals were present away from the injection top, thus showing greatly hindered precipitation (Figure 9f).

It is reported in literature that a change in precipitation pattern in this manner can be explained by the theory behind the competition between crystal growth and crystal nucleation [24,56]. Gandhi et al. [60] in their work unraveled that the nucleation of new crystals would face a competition with the process of crystal growth if the nucleation of new crystals is permitted over the growth of existing ones. In the case of high $Ca^{2+}$ and urea concentrations (1.0 M in current study), the cladding on grain surfaces by crystals is a result of crystal nucleation in new sites which is the initial dominant process. As the number of injections are repeated, higher and more localized pH rise around some bacterial cells take place, resulting in supersaturation due to the presence of more urea molecules and there is a greater tendency to accumulate precipitate over the initial smaller crystals rather than forming new crystals. This process will then lead to the formation of larger crystals which fill the void spaces and subsequently retard the flow of cementation solution. This has been further proved by Somani et al. [61] who stated that the average particle size of precipitate becomes larger as the carbonate concentration becomes higher.

**Figure 9.** SEM images of showing variation of crystal morphology for the samples treated under different bacteria and reactant concentrations: (**a–c**) 0.5 M cementation solution with bacteria injected twice; (**d–f**) 1.0 M cementation solution with bacteria injected only in the beginning.

In contrast to that, the nucleation of new crystals is more dominant than the growth of existing crystals when the carbonate concentration is lower [24]. The reduced shear stresses and availability of nutrients at particle contact points cause bacteria to prefer smaller surface features like grain contacts [62] whereas cells will also be distributed over grain surfaces depending the surface properties of grains [63]. Therefore, in this case, numerous smaller crystals will be first formed which will eventually be developed to a dense crystal layer with the repeated injection of cementation solution. From these results, it is apparent that uniformly distributed smaller crystals precipitated on grain surfaces and gaps between sand particles provide better strength and homogeneity rather than continuous heterogeneous precipitation of larger crystals. This is in line with the previous findings reported in literature. Okwadha and Li [41] stated that concentrations above 0.5 M of urea and $CaCl_2$ reduce the efficiency of MICP. Al Qabany et al. [56] and Ng et al. [12] also demonstrated that low concentrations of cementation solution yields higher specimen strengths. Moreover, the flow rate of solutions through the column also has a crucial effect on the precipitation process and thus should be maintained at a suitable rate for homogeneous strength improvement [64]. On the other hand, utilizing a different kind of bacteria or a different bacterial cell concentration would result in a precipitation pattern and crystal morphology different to what has been obtained in this study.

### 3.3.2. Effect of $Mg^{2+}$ Addition on Microbial Induced Carbonate Precipitation

Not much investigation have been carried out so far to investigate the effect of magnesium on the microbial inspired carbonate precipitation. Nevertheless, magnesium carbonate possesses similar cementing ability as calcium carbonate, which can significantly affect the desired outcomes of MICP treatment. Figure 10 illustrates the results of MICP test tube tests incorporating $Mg^{2+}$.

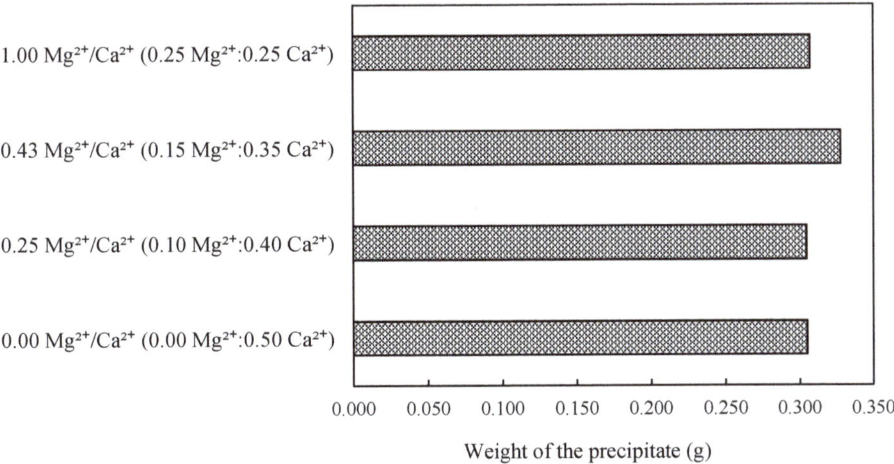


**Figure 10.** Variation of microbial induced carbonate precipitation under different $Mg^{2+}/Ca^{2+}$ molar ratios.

Although $Mg^{2+}/Ca^{2+}$ molar ratio was raised from 0 to 1, no remarkable change in the amount of precipitate was observed. However, the amount of carbonate precipitate slightly increased up to 0.43 $Mg^{2+}/Ca^{2+}$ ratio and then decreased as $Mg^{2+}$ was further raised until ratio became 1. Therefore, it was deduced that a $Mg^{2+}/Ca^{2+}$ ratio approximately equaled to 0.5 is the optimum condition in this case. Similarly, in a study by Sun et al. [38] demonstrated that magnesium carbonate has a lower productive rate than calcium carbonate resulting much lower magnesium carbonate precipitation than that of calcium carbonate in any specific condition. Further, Fukue et al. [65] claimed that the total amount of precipitation decreases with an increase in $Mg^{2+}/Ca^{2+}$ ratio from 0 to 2.3. This may be attributed to the higher solubility of magnesium carbonate which is almost one order of magnitude higher than its calcium counterpart and/or the possibility of partial hydrolysis of magnesium carbonate to generate magnesium hydroxide [38].

The size and morphology of the precipitated crystals has serious implications on the success of MICP treatment. Therefore, the microstructure and mineralogy of the precipitates from the aforementioned experimental cases were determined.

In the absence of $Mg^{2+}$, SEM imagery showed that the precipitate was mainly composed of entirely spherical crystals, some as large as 30–45 μm in diameter and agglomerations of relatively smaller crystals (Figure 11a). The XRD spectra revealed that the mineralogical composition was exclusively vaterite (Figure 12a), which is a metastable polymorph of $CaCO_3$. It is unstable, rapidly transforms into calcite (or aragonite) in aqueous solutions under room temperature, and forms necessarily in the presence of high supersaturation conditions [66]. The urea hydrolysis and resultant local alkalinity rise along with bacterial cells acting as nucleation sites might have developed a high supersaturation in the medium, giving rise to massive vaterite crystallization here. On the other hand, extracellular polymeric substances (EPS) play an important role in polymorph selection [67] and vaterite crystals have been abundantly reported with a high amount of accumulated EPS [68], which is possible in this case too. Moreover, vaterite generation is also preferred with the introduction of various organic matter to the aqueous system [69].

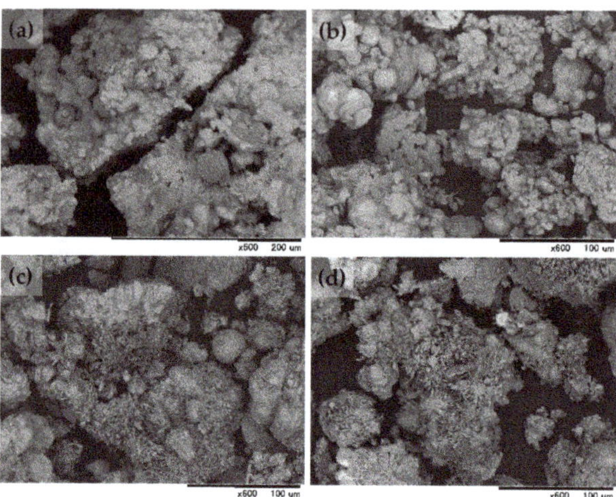

**Figure 11.** Scanning electron microscopy (SEM) images of microbial induced precipitate by *Sporosarcina* sp. with (**a**) $Mg^{2+}/Ca^{2+}$ = 0.00; (**b**) $Mg^{2+}/Ca^{2+}$ = 0.25; (**c**) $Mg^{2+}/Ca^{2+}$ = 0.43; (**d**) $Mg^{2+}/Ca^{2+}$ = 1.00.

As the $Mg^{2+}/Ca^{2+}$ molar ratio was stepped up to 0.25, the calcite (and magnesium calcite) peaks emerged at the expense of the proportion of vaterite peaks (Figure 12b). The corresponding change in SEM images was the smaller spherical crystals bursting and becoming fibrous (Figure 11b), as reported by Fukue et al. [65]. When the $Mg^{2+}/Ca^{2+}$ ratio is 0.43, aragonite was also spotted in the XRD spectra (Figure 12c). This is in consistent with the SEM images where the emergence of needle shaped crystals can be clearly seen (Figure 11c). Magnesium inducing the formation of needle/thorn shaped aragonite rather than calcite and vaterite in this manner, either abiotically or biotically, is a very well documented phenomenon in the literature [70–74]. The change of the crystals from spherical to a needle-like shape is a result of the inhibition of crystal growth in the a and b axes while allowing the same along the c axis in the presence of magnesium [65]. Substitution of $Mg^{2+}$ until the $Mg^{2+}/Ca^{2+}$ ratio became 1.0 increased the intensity of aragonite peaks further showing a mixture of other peaks, including calcite and magnesium calcite (Figure 12d). Subsequently, vaterite peaks disappeared. The corresponding change in the SEM micrographs is the substantial appearance of the needle shaped crystals fanning out into the space in all directions (Figure 11d).

These observations clearly revealed that although increments of $Mg^{2+}$ did not affect the quantity of bio-induced precipitation to a great deal, the morphology of the crystals was far more sensitive to them.

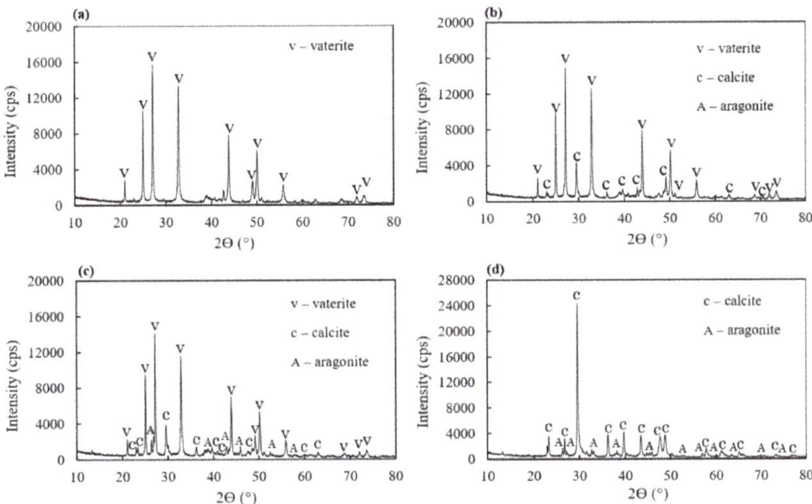

**Figure 12.** X-ray diffraction (XRD) spectra of the microbial induced precipitate by *Sporosarcina* sp. with
(**a**) $Mg^{2+}/Ca^{2+} = 0.00$; (**b**) $Mg^{2+}/Ca^{2+} = 0.25$; (**c**) $Mg^{2+}/Ca^{2+} = 0.43$; (**d**) $Mg^{2+}/Ca^{2+} = 1.00$.

The precipitation results alone are not adequate to predict the effectiveness of MICP for beach
sand solidification in the presence of magnesium. Therefore, syringe sand solidification tests were
conducted and estimated UCS are displayed in Figure 13.

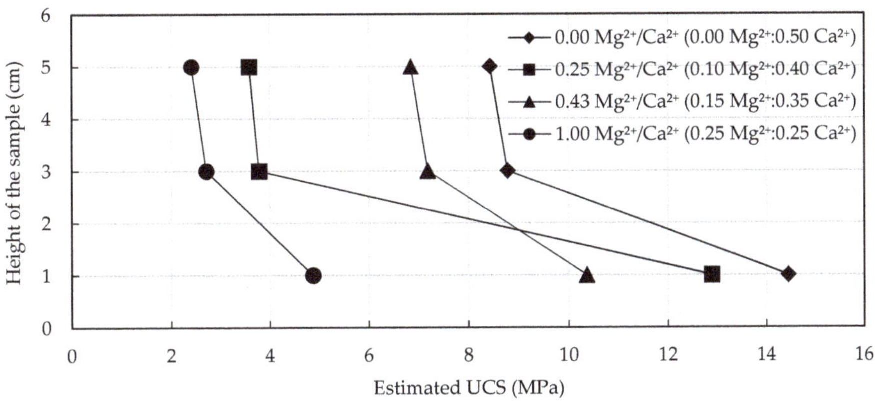

**Figure 13.** Variation of local strength of MICP treated sand for different $Mg^{2+}/Ca^{2+}$ molar ratios.

Overall, it can be seen that the strength improvement by MICP treatment is considerably impaired
in the presence of magnesium, and specimens without $Mg^{2+}$ show the highest strength gain. Evidently,
the specimens treated with an $Mg^{2+}/Ca^{2+}$ ratio of 0.43 exhibit better strength than those with an
$Mg^{2+}/Ca^{2+}$ ratio of 0.25. This can be partly supported by the results of precipitation tests where the
highest amount of precipitate was obtained for the $Mg^{2+}/Ca^{2+} = 0.43$ condition.

SEM microscopic observations were conducted in order to ascertain the effect of crystal morphology
on strength enhancement. It is clear that the precipitated crystal morphology changed from
rhombohedral to spherical and then to needle shape when the $Mg^{2+}/Ca^{2+}$ ratio gradually advanced
from 0 to 1 (Figure 14). In the absence of $Mg^{2+}$, the crystal morphology was exclusively rhombohedral,
the typical crystalline form of calcite (Figure 14a). The grain surfaces were fully covered with crystals

of average size 10 μm and matrix was supported by relatively larger crystals (Figure 14b). The crystals were very well interweaved and crosslinked at the grain contact points thereby effectively bonding the loose sand particles together. When $Mg^{2+}/Ca^{2+}$ ratio was 0.25 only, the crystals showed slight alterations from rhombohedral form whereas contact cementing and matrix supporting seemed not as significant as in previous case (Figure 14c). As $Mg^{2+}/Ca^{2+}$ ratio was increased further to 0.43, crystals became spherical and accumulated well together at contact points (Figure 14d). Fibrous needle shaped crystals bundled together were also abundantly seen on grain surfaces as well as on contact cements. This pattern of crystal morphology resulted in a stronger bond between grains thus yielding compressive strengths closer to that without $Mg^{2+}$.

**Figure 14.** SEM images of variation in crystal morphology with different $Mg^{2+}/Ca^{2+}$ molar ratios: (**a**,**b**) $Mg^{2+}/Ca^{2+}$ = 0.00 and its closer view; (**c**) $Mg^{2+}/Ca^{2+}$ = 0.25; (**d**) $Mg^{2+}/Ca^{2+}$ = 0.43; (**e**,**f**); $Mg^{2+}/Ca^{2+}$ = 1.00 and its closer view.

Stepping up the $Mg^{2+}/Ca^2$ ratio from 0.43 to 1.00 drastically changed the crystal morphology of the biocement. Grains were slightly cladded by small crystals while pore spaces were completely occupied by needle shaped crystals, some as long as 100–200 μm (Figure 14e,f). Apparently, these thin crystals, individually fanning out in all directions over the void spaces, had a very low binding capacity compared to the rhombohedral/spherical crystals accumulated together at grain contacts. Further, thin needle shaped crystals can be easily subjected to destruction on application of compressive forces compared to the rhombohedral or spherical crystal aggregates. These results are in good agreement with that of Gawwad et al. [73] who showed that the occurrence of needle cement in the presence of higher concentrations of magnesium deteriorates the compressive strength of the bio-mortar prepared by the common ureolytic bacterium *Sporosarcina pasteurii*. However, the majority of the studies done to address this aspect has revealed that magnesium addition enables the biocement to have a higher strength, lower porosity, and better resistance to acids compared to control samples in absence of magnesium [19,38,65,70,75,76]. The general understanding from the above studies is that the inherent properties of aragonite, such as higher specific gravity and Mohs hardness than those of calcite and morphology which provides a denser, compact arrangement, can be the possible reason for better results. The exact reasoning to this discrepancy in results are not fully discovered yet and need more investigation.

Another interesting observation that should be noted here is the local strength variation along the sand column. Unlike the results presented under the Section 3.3.1, the estimated compressive strength increases with sand column depth. This can be attributed to the differences in particle size gradation of the two sands (sand 1 and sand 2) used. "Sand 1" is a well graded sand with a considerable amount of fines percentage while "sand 2" is a comparatively coarser uniformly graded sand. As the sand becomes finer and well graded, the sand becomes densely packed, void spaces become narrower and fewer, and the sand offers more particle-to-particle contact. The movement of bacteria through the sand column is obviously through the pore throats between grains either by self-propelled movement or by passive diffusion, and the geometric compatibility between the size of bacteria and that of the pore throat is a crucial factor [12]. In a dense particle arrangement like in sand 1, bacteria may be filtered out and cell concentration may reduce from top to bottom, whereas in a loose particle arrangement like in sand 2, bacteria may be easily flushed out with daily dose of cementation solution and retained at the bottom [44,50]. This can lead to a better cementation on top in sand 1 and on bottom for sand 2. This dictates that distribution of bacteria is a critical factor in the process.

Many researchers that have attempted to exploit the effect of grain size and gradation have presented similar results [58,77–79]. However, Dhami et al. [58] stated that availability of oxygen plays the governing role rather than particle size distribution, thus efficacy of MICP is reduced with depth irrespective of the grain sizes. The pertinent reason for contrasting results may be that the *Sporosarcina* sp. strain used in this study has a greater tolerance to oxygen depletion than the *Bacillus megaterium* strain used in the aforementioned study, as the column length is only in centimeter scale.

## 4. Conclusions and Implications for the Target Application

This paper intends to report the results of a preliminary study carried out to investigate the potential of implementing microbial induced carbonate precipitation (MICP) as a novel approach towards stabilizing loose beach sand in Sri Lanka to minimize coastal erosion. This is the first study to address this aspect in the aforementioned region. After conducting a series of laboratory tests, the following conclusions were drawn.

Although ureolytic bacteria are ubiquitous in nature, not all have the required potential for MICP applications. The isolated native bacterium belonging to *Sporosarcina* sp. showed sufficient feasibility for the intended application. It is a moderately alkalo-tolerant bacterium, showing maximum growth and urease activity at an ambient temperature of 25 °C which is compatible with the tropical climate prevailing in the target region.

Laboratory scale (syringe) sand solidification test results depicted that a batch treatment of bacterial culture injected twice (in the beginning and after 7 days) with a daily injection of 0.5 M cementation solution yielded sufficiently homogeneous strength improvement after a treatment period of 14 days under saturated conditions. The residual $Ca^{2+}$ and pH measurements of the effluent draining out from the bottom provided substantial information about the rate of urea hydrolysis and eventual calcium carbonate precipitation and can be used as good indication of the state of MICP.

Local strength analysis of two types of natural beach sand solidified by MICP exhibited that particle size and gradation, bacterial cell concentration, and the concentration of reactants have a great deal of importance on the degree and homogeneity of the strength enhancement. A precipitation pattern of uniformly distributed crystals cladding grain surfaces along with agglomerated crystal clusters at the particle contact points showed high efficiency in improving the compressive strength of biocemented sands than randomly distributed larger crystals which may result under high reactant concentrations.

Introducing magnesium into the MICP process deemed to negatively affect the desired engineering properties of the treated specimens. At the extreme, excessively high concentrations of magnesium where the $Mg^{2+}/Ca^2$ molar ratio was 1.0, resulted in a significant reduction in the UCS. It is clearly deduced that the strength decline is mainly due to the change in crystal morphology (rather than the quantity of precipitate) to a needle shaped crystal form with lower bonding capacity than that of the well-interweaved and cross-linked rhombohedral crystals produced in the absence of magnesium. However, the effect will be negligible in the target environment where naturally available $Mg^{2+}$ is considerably lower compared to the concentration of $Ca^{2+}$ supplied with the cementation solution.

On the other hand, the sand columns in this study were cemented under one-dimensional vertical flow of solutions where opted pathway of solutions were confined by syringe walls. In a real field application, the flow will be three-dimensional where flow dynamics will be dependent on numerous factors including the characteristics of the solution, sand matrix and the groundwater level.

Finally, there is still need for further experiments, under laboratory and natural conditions, in order to scale up the process and bring them to practical use. The ultimate goal of this part of the current research study was to evaluate the initial feasibility for MICP as an ecofriendly alternative approach towards beach sand stabilization and to determine the preliminary optimum conditions, making this technique one step closer to in-situ application.

**Supplementary Materials:** The following are available online at http://www.mdpi.com/2076-3417/9/15/3201/s1, Figure S1: Location map and study area, Figure S2: Particle size distribution of natural beach sand, Figure S3: Experimental set up for the syringe sand solidification test, Figure S4: Bacterial cell concentration of isolated strains.

**Author Contributions:** P.G.N.N. performed laboratory experiments, analysis, interpretation of data and original draft preparation; A.B.N.D. and K.N. provided technical assistance and reviewed and edited the manuscript; S.K. did primary supervision and helped in analyzing, methodical guidance, critical reviewing and approval of final version to be submitted.

**Funding:** This research was partially funded by JSPS KAKENHI, grant numbers JP16H04404 and JP19H02229, Japan. The authors gratefully acknowledge the support.

**Conflicts of Interest:** The authors declare no conflict of interest.

## References

1. DeJong, J.T.; Mortensen, B.M.; Martinez, B.C.; Nelson, D.C. Bio-mediated soil improvement. *Ecol. Eng.* **2010**, *36*, 197–210. [CrossRef]

2. Khan, M.N.H.; Shimazaki, S.; Kawasaki, S. Coral Sand Solidification Test Through Microbial Calcium Carbonate Precipitation Using *Pararhodobacter* sp. *Int. J. GEOMATE* **2016**, *11*, 2665–2670.

3. Gowthaman, S.; Mitsuyama, S.; Nakashima, K.; Komatsu, M.; Kawasaki, S. Microbial Induced Slope Surface Stabilization Using Industrial-Grade Chemicals: A Preliminary Laboratory Study. *Int. J. GEOMATE* **2019**, *17*, 110–116. [CrossRef]

4. Van Paassen, L.A.; Ghose, R.; van der Linden, T.J.M.; van der Star, W.R.L.; van Loosdrecht, M.C.M. Quantifying Biomediated Ground Improvement by Ureolysis: Large-Scale Biogrout Experiment. *J. Geotech. Geoenviron. Eng.* **2010**, *136*, 1721–1728. [CrossRef]

5. Gomez, M.G.; Anderson, C.M.; Graddy, C.M.R.; DeJong, J.T.; Nelson, D.C.; Ginn, T.R. Large-Scale Comparison of Bioaugmentation and Biostimulation Approaches for Biocementation of Sands. *J. Geotech. Geoenviron. Eng.* **2016**, *143*, 04016124. [CrossRef]

6. Canakci, H.; Sidik, W.; Halil Kilic, I. Effect of bacterial calcium carbonate precipitation on compressibility and shear strength of organic soil. *Soils Found.* **2015**, *55*, 1211–1221. [CrossRef]

7. Cheng, L.; Cord-Ruwisch, R. In situ soil cementation with ureolytic bacteria by surface percolation. *Ecol. Eng.* **2012**, *42*, 64–72. [CrossRef]

8. Zhao, Y.; Yao, J.; Yuan, Z.; Wang, T.; Zhang, Y.; Wang, F. Bioremediation of Cd by strain GZ-22 isolated from mine soil based on biosorption and microbially induced carbonate precipitation. *Environ. Sci. Pollut. Res.* **2017**, *24*, 372–380. [CrossRef]

9. Wu, J.; Wang, X.B.; Wang, H.F.; Zeng, R.J. Microbially induced calcium carbonate precipitation driven by ureolysis to enhance oil recovery. *RSC Adv.* **2017**, *7*, 37382–37391. [CrossRef]

10. Okyay, T.O.; Nguyen, H.N.; Castro, S.L.; Rodrigues, D.F. CO$_2$ sequestration by ureolytic microbial consortia through microbially-induced calcite precipitation. *Sci. Total Environ.* **2016**, *572*, 671–680. [CrossRef]

11. Mwandira, W.; Nakashima, K.; Kawasaki, S.; Ito, M.; Sato, T.; Igarashi, T.; Banda, K.; Chirwa, M.; Nyambe, I.; Nakayama, S.; et al. Efficacy of biocementation of lead mine waste from the Kabwe Mine site evaluated using *Pararhodobacter* sp. *Environ. Sci. Pollut. Res.* **2019**, *26*, 15653–15664. [CrossRef] [PubMed]

12. Ng, W.S.; Lee, M.-L.; Hii, S.-L. An Overview of the Factors Affecting Microbial-Induced Calcite Precipitation and its Potential Application in Soil Improvement. *Int. J. Civ. Environ. Eng.* **2012**, *6*, 188–194.

13. Stocks-Fischer, S.; Galinat, J.K.; Bang, S.S. Microbiological precipitation of CaCO$_3$. *Soil Biol. Biochem.* **1999**, *31*, 1563–1571. [CrossRef]

14. Harkes, M.P.; van Paassen, L.A.; Booster, J.L.; Whiffin, V.S.; van Loosdrecht, M.C.M. Fixation and distribution of bacterial activity in sand to induce carbonate precipitation for ground reinforcement. *Ecol. Eng.* **2010**, *36*, 112–117. [CrossRef]

15. Cortés, P.; Escrig, C.; Bernat-Maso, E.; Gil, L.; Barbé, J. Effect of *Sporosarcina pasteurii* on the strength properties of compressed earth specimens. *Mater. Construcción* **2018**, *68*, 143.

16. Whiffin, V.S.; van Paassen, L.A.; Harkes, M.P. Microbial carbonate precipitation as a soil improvement technique. *Geomicrobiol. J.* **2007**, *24*, 417–423. [CrossRef]

17. Temmerman, S.; Meire, P.; Bouma, T.J.; Herman, P.M.J.; Ysebaert, T.; De Vriend, H.J. Ecosystem-based coastal defence in the face of global change. *Nature* **2013**, *504*, 79–83. [CrossRef]

18. Ten Voorde, M.; do Carmo, J.S.A.; Neves, M.G. Designing a Preliminary Multifunctional Artificial Reef to Protect the Portuguese Coast. *J. Coast. Res.* **2009**, *251*, 69–79. [CrossRef]

19. Cheng, L.; Shahin, M.A.; Cord-Ruwisch, R. Bio-cementation of sandy soil using microbially induced carbonate precipitation for marine environments. *Géotechnique* **2014**, *64*, 1010–1013. [CrossRef]

20. Salifu, E.; MacLachlan, E.; Iyer, K.R.; Knapp, C.W.; Tarantino, A. Application of microbially induced calcite precipitation in erosion mitigation and stabilisation of sandy soil foreshore slopes: A preliminary investigation. *Eng. Geol.* **2016**, *201*, 96–105. [CrossRef]

21. Chu, J.; Ivanov, V.; Chenghong, G.; Naeimi, M.; Tkalich, P. Microbial geotechnical engineering for disaster mitigation and coastal management. *Proc. Earthq. Tsunami* **2009**, 1–6.

22. Shanahan, C.; Montoya, B.M. *Strengthening Coastal Sand Dunes Using Microbial-Induced Calcite Precipitation*; Geo-Congress 2014 Technical Papers; American Society of Civil Engineers (ASCE): Reston, VA, USA, 2014; pp. 1683–1692.

23. Van der Ruyt, M.; van der Zon, W. Biological in situ reinforcement of sand in near-shore areas. *Proc. Inst. Civ. Eng. Geotech. Eng.* **2009**, *162*, 81–83. [CrossRef]

24. Cheng, L.; Shahin, M.A.; Mujah, D. Influence of Key Environmental Conditions on Microbially Induced Cementation for Soil Stabilization. *J. Geotech. Geoenviron. Eng.* **2016**, *143*, 04016083-1–04016083-11. [CrossRef]

25. Gat, D.; Tsesarsky, M.; Shamir, D.; Ronen, Z. Accelerated microbial-induced CaCO$_3$ precipitation in a defined coculture of ureolytic and non-ureolytic bacteria. *Biogeosciences* **2014**, *11*, 2561–2569. [CrossRef]

26. De Muynck, W.; De Belie, N.; Verstraete, W. Microbial carbonate precipitation in construction materials: A review. *Ecol. Eng.* **2010**, *36*, 118–136. [CrossRef]

27. Lin, H.; Suleiman, M.T.; Brown, D.G.; Kavazanjian, E. Mechanical Behavior of Sands Treated by Microbially Induced Carbonate Precipitation. *J. Geotech. Geoenviron. Eng.* **2015**, *142*, 04015066. [CrossRef]

28. Gat, D.; Ronen, Z.; Tsesarsky, M. Soil Bacteria Population Dynamics Following Stimulation for Ureolytic Microbial-Induced CaCO$_3$ Precipitation. *Environ. Sci. Technol.* **2016**, *50*, 616–624. [CrossRef] [PubMed]

29. Tobler, D.J.; Cuthbert, M.O.; Greswell, R.B.; Riley, M.S.; Renshaw, J.C.; Handley-Sidhu, S.; Phoenix, V.R. Comparison of rates of ureolysis between *Sporosarcina pasteurii* and an indigenous groundwater community under conditions required to precipitate large volumes of calcite. *Geochim. Cosmochim. Acta* **2011**, *75*, 3290–3301. [CrossRef]

30. Dhami, N.K.; Alsubhi, W.R.; Watkin, E.; Mukherjee, A. Bacterial community dynamics and biocement formation during stimulation and augmentation: Implications for soil consolidation. *Front. Microbiol.* **2017**, *8*, 1267. [CrossRef]

31. Wenderoth, D.F.; Rosenbrock, P.; Abraham, W.R.; Pieper, D.H.; Höfle, M.G. Bacterial community dynamics during biostimulation and bioaugmentation experiments aiming at chlorobenzene degradation in groundwater. *Microb. Ecol.* **2003**, *46*, 161–176. [CrossRef]

32. Nawarathna, T.H.K.; Nakashima, K.; Fujita, M.; Takatsu, M.; Kawasaki, S. Effects of Cationic Polypeptide on $CaCO_3$ Crystallization and Sand Solidification by Microbial-Induced Carbonate Precipitation. *ACS Sustain. Chem. Eng.* **2018**, *6*, 10315–10322. [CrossRef]

33. Graddy, C.M.R.; Gomez, M.G.; Kline, L.M.; Morrill, S.R.; Dejong, J.T.; Nelson, D.C. Diversity of *Sporosarcina* -like Bacterial Strains Obtained from Meter-Scale Augmented and Stimulated Biocementation Experiments. *Environ. Sci. Technol.* **2018**, *52*, 3997–4005. [CrossRef] [PubMed]

34. Sun, Y.; Zhao, Q.; Zhi, D.; Wang, Z.; Wang, Y.; Xie, Q.; Wu, Z.; Wang, X.; Li, Y.; Yu, L.; et al. *Sporosarcina terrae* sp. Nov., isolated from orchard soil. *Int. J. Syst. Evol. Microbiol.* **2017**, *67*, 2104–2108. [CrossRef] [PubMed]

35. Dick, J.; De Windt, W.; De Graef, B.; Saveyn, H.; Van Der Meeren, P.; De Belie, N.; Verstraete, W. Bio-deposition of a calcium carbonate layer on degraded limestone by *Bacillus* species. *Biodegradation* **2006**, *17*, 357–367. [CrossRef] [PubMed]

36. Achal, V.; Mukherjee, A.; Basu, P.C.; Reddy, M.S. Strain improvement of *Sporosarcina pasteurii* for enhanced urease and calcite production. *J. Ind. Microbiol. Biotechnol.* **2009**, *36*, 981–988. [CrossRef] [PubMed]

37. Hammad, I.A.; Talkhan, F.N.; Zoheir, A.E. Urease activity and induction of calcium carbonate precipitation by *Sporosarcina pasteurii* NCIMB 8841. *J. Appl. Sci. Res.* **2013**, *9*, 1525–1533.

38. Sun, X.; Miao, L.; Wu, L.; Wang, C. Study of magnesium precipitation based on biocementation. *Mar. Georesour. Geotechnol.* **2019**. [CrossRef]

39. Kang, C.H.; Choi, J.H.; Noh, J.G.; Kwak, D.Y.; Han, S.H.; So, J.S. Microbially Induced Calcite Precipitation-based Sequestration of Strontium by *Sporosarcina pasteurii* WJ-2. *Appl. Biochem. Biotechnol.* **2014**, *174*, 2482–2491. [CrossRef] [PubMed]

40. Bang, S.S.; Galinat, J.K.; Ramakrishnan, V. Calcite precipitation induced by polyurethane-immobilized *Bacillus pasteurii*. *Enzyme Microb. Technol.* **2001**, *28*, 404–409. [CrossRef]

41. Okwadha, G.D.O.; Li, J. Optimum conditions for microbial carbonate precipitation. *Chemosphere* **2010**, *81*, 1143–1148. [CrossRef]

42. Fujita, Y.; Grant Ferris, F.; Daniel Lawson, R.; Colwell, F.S.; Smith, R.W. Calcium carbonate precipitation by ureolytic subsurface bacteria. *Geomicrobiol. J.* **2000**, *17*, 305–318. [CrossRef]

43. Anbu, P.; Kang, C.H.; Shin, Y.J.; So, J.S. Formations of calcium carbonate minerals by bacteria and its multiple applications. *Springerplus* **2016**, *5*, 1–26. [CrossRef] [PubMed]

44. Rebata-Landa, V. Microbial Activity in Sediments: Effects on Soil Behavior. Ph.D. Thesis, Georgia Institute of Technology, Atlanta, GA, USA, 2007.

45. Dhami, N.K.; Reddy, M.S.; Mukherjee, A. Synergistic role of bacterial urease and carbonic anhydrase in carbonate mineralization. *Appl. Biochem. Biotechnol.* **2014**, *172*, 2552–2561. [CrossRef] [PubMed]

46. Sahrawat, K.L. Effects of temperature and moisture on urease activity in semi-arid tropical soils. *Plant Soil* **1984**, *8*, 401–408. [CrossRef]

47. Liang, Z.P.; Feng, Y.Q.; Meng, S.X.; Liang, Z.Y. Preparation and properties of urease immobilized onto Glutaraldehyde cross-linked Chitosan beads. *Chin. Chem. Lett.* **2005**, *16*, 135–138.

48. Chen, Y.Y.M.; Anne Clancy, K.; Burne, R.A. *Streptococcus salivarius* urease: Genetic and biochemical characterization and expression in a dental plaque streptococcus. *Infect. Immun.* **1996**, *64*, 585–592.

49. Fujita, M.; Nakashima, K.; Achal, V.; Kawasaki, S. Whole-cell evaluation of urease activity of *Pararhodobacter* sp. isolated from peripheral beachrock. *Biochem. Eng. J.* **2017**, *124*, 1–5. [CrossRef]

50. Van Paassen, L.A. Biogrout: Ground Improvement by Microbially Induced Carbonate Precipitation. Ph.D. Thesis, Delft University of Technology, Delft, The Netherlands, 2009.

51. Mobley, H.L.T.; Island, M.D.; Hausinger, R.P. Molecular biology of microbial ureases. *Microbiol. Rev.* **1995**, *59*, 451–480. [PubMed]

52. Ciurli, S.; Marzadori, C.; Benini, S.; Deiana, S.; Gessa, C. Urease from the soil bacterium *Bacillus pasteurii*: Immobilization on Ca- polygalacturonate. *Soil Biol. Biochem.* **1996**, *28*, 811–817. [CrossRef]

53. Mahadevan, S.; Sauer, F.D.; Erfle, J.D. Purification and properties of urease from bovine rumen. *Biochem. J.* **2015**, *163*, 495–501. [CrossRef]

54. Ng, W.S.; Lee, L.M.; Khun, T.C.; Ling, H.S. Factors Affecting Improvement in Engineering Properties of Residual Soil through Microbial-Induced Calcite Precipitation. *J. Geotech. Geoenviron. Eng.* **2014**, *140*, 04014006.

55. Zhu, T.; Dittrich, M. Carbonate Precipitation through Microbial Activities in Natural Environment, and Their Potential in Biotechnology: A Review. *Front. Bioeng. Biotechnol.* **2016**, *4*, 1–21. [CrossRef] [PubMed]

56. Al Qabany, A.; Soga, K.; Santamarina, C. Factors Affecting Efficiency of Microbially Induced Calcite Precipitation. *J. Geotech. Geoenviron. Eng.* **2011**, *138*, 992–1001. [CrossRef]

57. Li, W.; Liu, L.P.; Zhou, P.P.; Cao, L.; Yu, L.J.; Jiang, S.Y. Calcite precipitation induced by bacteria and bacterially produced carbonic anhydrase. *Curr. Sci.* **2011**, *100*, 502–508.

58. Dhami, N.K.; Reddy, M.S.; Mukherjee, A. Significant indicators for biomineralisation in sand of varying grain sizes. *Constr. Build. Mater.* **2016**, *104*, 198–207. [CrossRef]

59. Gorospe, C.M.; Han, S.H.; Kim, S.G.; Park, J.Y.; Kang, C.H.; Jeong, J.H.; So, J.S. Effects of different calcium salts on calcium carbonate crystal formation by *Sporosarcina pasteurii* KCTC 3558. *Biotechnol. Bioprocess Eng.* **2013**, *18*, 903–908. [CrossRef]

60. Gandhi, K.S.; Kumar, R.; Ramkrishna, D. Some Basic Aspects of Reaction Engineering of Precipitation Processes. *Ind. Eng. Chem. Res.* **1995**, *34*, 3223–3230. [CrossRef]

61. Somani, R.S.; Patel, K.S.; Mehta, A.R.; Jasra, R.V. Examination of the polymorphs and particle size of calcium carbonate precipitated using still effluent (i.e., $CaCl_2$ + NaCl solution) of soda ash manufacturing process. *Ind. Eng. Chem. Res.* **2006**, *45*, 5223–5230. [CrossRef]

62. Mujah, D.; Shahin, M.A.; Cheng, L. State-of-the-Art Review of Biocementation by Microbially Induced Calcite Precipitation (MICP) for Soil Stabilization. *Geomicrobiol. J.* **2017**, *34*, 524–537. [CrossRef]

63. Chu, J.; Ivanov, V.; Naeimi, M.; Stabnikov, V.; Liu, H.L. Optimization of calcium-based bioclogging and biocementation of sand. *Acta Geotech.* **2014**, *9*, 277–285. [CrossRef]

64. Mortensen, B.M.; Haber, M.J.; DeJong, J.T.; Caslake, L.F.; Nelson, D.C. Effects of environmental factors on microbial induced calcium carbonate precipitation. *J. Appl. Microbiol.* **2011**, *111*, 338–349. [CrossRef] [PubMed]

65. Fukue, M.; Ono, S.-I.; Sato, Y. Cementation of Sands Due To Microbiologically-Induced Carbonate Precipitation. *Soils Found.* **2011**, *51*, 83–93. [CrossRef]

66. Spanos, N.; Koutsoukos, P.G. The transformation of vaterite to calcite: Effect of the conditions of the solutions in contact with the mineral phase. *J. Cryst. Growth* **1998**, *191*, 783–790. [CrossRef]

67. Bains, A.; Dhami, N.K.; Mukherjee, A.; Reddy, M.S. Influence of Exopolymeric Materials on Bacterially Induced Mineralization of Carbonates. *Appl. Biochem. Biotechnol.* **2015**, *175*, 3531–3541. [CrossRef] [PubMed]

68. Gowthaman, S.; Mitsuyama, S.; Nakashima, K.; Komatsu, M.; Kawasaki, S. Biogeotechnical approach for slope soil stabilization using locally isolated bacteria and inexpensive low-grade chemicals: A feasibility study on Hokkaido expressway soil, Japan. *Soils Found.* **2019**, *59*, 484–499. [CrossRef]

69. Nawarathna, T.H.K.; Nakashima, K.; Kawasaki, S. Chitosan enhances calcium carbonate precipitation and solidification mediated by bacteria. *Int. J. Biol. Macromol.* **2019**, *133*, 867–874. [CrossRef] [PubMed]

70. Putra, H.; Kinoshita, N.; Yasuhara, H.; Lu, C.-W.; Neupane, D. Effect of Magnesium as Substitute Material in Enzyme-Mediated Calcite Precipitation for Soil-Improvement Technique. *Front. Bioeng. Biotechnol.* **2016**, *4*, 37. [CrossRef] [PubMed]

71. Mejri, W.; Korchef, A.; Tlili, M.; Ben Amor, M. Effects of temperature on precipitation kinetics and microstructure of calcium carbonate in the presence of magnesium and sulphate ions. *Desalin. Water Treat.* **2014**, *52*, 4863–4870. [CrossRef]

72. Oomori, T.; Kitano, Y. The effect of Magnesium ions on the polymorphic crystallization of calcium carbonate. *Bull. Coll. Sci. Univ. Ryukyus* **1985**, *39*, 57–62.

73. Gawwad, H.A.A.; Mohamed, S.A.E.A.; Mohammed, S.A. Impact of magnesium chloride on the mechanical properties of innovative bio-mortar. *Mater. Lett.* **2016**, *178*, 39–43. [CrossRef]

74. Putra, H.; Yasuhara, H.; Kinoshita, N.; Hirata, A. Optimization of Enzyme-Mediated Calcite Precipitation as a Soil-Improvement Technique: The Effect of Aragonite and Gypsum on the Mechanical Properties of Treated Sand. *Crystals* **2017**, *7*, 59. [CrossRef]

75. Sun, X.; Miao, L. The Comparison of Microbiologically-Induced Calcium Carbonate Precipitation and Magnesium Carbonate Precipitation. In *Proceedings of the 8th International Congress on Environmental Geotechnics*; Springer Nature Singapore Pte Ltd.: Singapore, 2019; Volume 3, pp. 302–308.

76. Rong, H.; Qian, C.X.; Li, L.Z. Influence of Magnesium Additive on Mechanical Properties of Microbe Cementitious Materials. *Mater. Sci. Forum* **2013**, *743–744*, 275–279. [CrossRef]

77. Mahawish, A.; Bouazza, A.; Gates, W.P. Effect of particle size distribution on the bio-cementation of coarse aggregates. *Acta Geotech.* **2018**, *13*, 1019–1025. [CrossRef]

78. Sharma, A.; Ramkrishnan, R. Study on effect of Microbial Induced Calcite Precipitates on strength of fine grained soils. *Perspect. Sci.* **2016**, *8*, 198–202. [CrossRef]

79. Deng, W.; Wang, Y. Investigating the Factors Affecting the Properties of Coral Sand Treated with Microbially Induced Calcite Precipitation. *Adv. Civ. Eng.* **2018**, *2018*, 1–6. [CrossRef]

*Article*

# Feasibility Study of Native Ureolytic Bacteria for Biocementation Towards Coastal Erosion Protection by MICP Method

**Md Al Imran [1], Shuya Kimura [1], Kazunori Nakashima [2], Niki Evelpidou [3] and Satoru Kawasaki [2,***

[1] Division of Sustainable Resources Engineering, Graduate School of Engineering, Hokkaido University, Sapporo 060-8628, Japan
[2] Division of Sustainable Resources Engineering, Faculty of Engineering, Hokkaido University, Sapporo 060-8628, Japan
[3] School of Science, Faculty of Geology and Geoenvironment, National and Kapodistrian University of Athens, 15784 Athens, Greece
* Correspondence: kawasaki@geo-er.eng.hokudai.ac.jp; Tel.: +81-11-706-6318

Received: 17 September 2019; Accepted: 16 October 2019; Published: 21 October 2019

**Abstract:** In recent years, traditional material for coastal erosion protection has become very expensive and not sustainable and eco-friendly for the long term. As an alternative countermeasure, this study focused on a sustainable biological ground improvement technique that can be utilized as an option for improving the mechanical and geotechnical engineering properties of soil by the microbially induced carbonate precipitation (MICP) technique considering native ureolytic bacteria. To protect coastal erosion, an innovative and sustainable strategy was proposed in this study by means of combing geotube and the MICP method. For a successful sand solidification, the urease activity, environmental factors, urease distribution, and calcite precipitation trend, among others, have been investigated using the isolated native strains. Our results revealed that urease activity of the identified strains denoted as G1 (*Micrococcus* sp.), G2 (*Pseudoalteromonas* sp.), and G3 (*Virgibacillus* sp.) relied on environment-specific parameters and, additionally, urease was not discharged in the culture solution but would discharge in and/or on the bacterial cell, and the fluid of the cells showed urease activity. Moreover, we successfully obtained solidified sand bearing UCS (Unconfined Compressive Strength) up to 1.8 MPa. We also proposed a novel sustainable approach for field implementation in a combination of geotube and MICP for coastal erosion protection that is cheaper, energy-saving, eco-friendly, and sustainable for Mediterranean countries, as well as for bio-mediated soil improvement.

**Keywords:** microbially induced carbonate precipitation; ureolytic bacteria; urease activity; biomineralization; coastal erosion protection; artificial beachrock

---

## 1. Introduction

At present, coastal erosion is a major problem all over the world. Traditional measures consisting of hard and soft defense protection systems are adopted to protect from this disintegration. All of these coastal design and structures have been identified as being expensive to build and sustain because of the scarcity of materials, energy, time, cost, and environmental concern. Therefore, sustainable and eco-friendly materials and systems are in demand for the protection against coastal erosion. Countermeasures against coastal erosion protection have been studied by many researchers [1], including hard and soft structures focusing on urea degradation bacteria. However, all these methods have an adverse effect on the surrounding landscape, environment, and ecosystem [2]; moreover, eco-sustainability is another major concern.

As an alternative method, earlier studies used the microbially induced carbonate precipitation (MICP) technique focusing on ureolytic bacteria within the field of geotechnology and civil engineering [3]. Among them, T. Danjo and S. Kawasaki conducted several studies in Ishikawa and Okinawa, Japan focusing on beachrock formation mechanism [4,5]. They found significant information to protect against coastal erosion using the MICP method, which they called "artificial beachrock". Moreover, for coastal erosion protection, the attention of using hard and soft structures, such as geo-tube, beach drainage system, vegetation, sand fencing, artificial reefs, and green belts, are dominating with progressive knowledge [6]. Recently, geo-tube technology has converted from being a substitute construction technique to a key solution of choice in a more advanced way for preventing water logging conditions, as well as coastal erosion protection [7]. The traditional geotextile tube technology for coastal erosion protection is not sustainable because of the penetration of UV (Ultraviolet) rays that could accelerate to damage the geosynthetic product very easily; moreover, it is rather difficult to recover or repair, which would be considered as costly [8]. Therefore, this technology needs to be improved and, hence, we proposed a novel approach to protect coastal erosion through a combination of MICP and geotube, which is more eco-friendly, cost-effective, and more sustainable.

There are various metabolic pathways leading to MICP, and urea hydrolysis is the most prominent for the formation of beachrocks [8,9]. The reactions occurring in this mechanism are represented by the following equations (Equations (1)–(3)), wherein urea hydrolysis occurs by urease enzyme and produces ammonium ion and carbamate, mentioned in Equation (1), and subsequently, the carbamate ion is hydrolyzed to form an additional ammonia ion and bi-carbonate, shown in Equation (2) [10,11].

$$CO(NH_2)_2 + H_2O \overset{Urease}{\rightarrow} H_2NCOO^- + NH_4^+, \tag{1}$$

$$H_2NCOO^- + H_2O \rightarrow HCO_3^- + NH_3. \tag{2}$$

Then, calcium carbonate is formed and finally precipitated, which plays a major role for binding the sand particles. In the presence of $Ca^{2+}$ ions, calcium carbonate is then formed and precipitated, which is effective for the binding of sand particles and for the filling of micro-space:

$$Ca^{2+} + HCO_3^- + NH_3 \rightarrow CaCO_3 \downarrow + NH_4^+. \tag{3}$$

In the MICP method, precipitation of carbonate minerals is an important factor that occurs through extracellular or intracellular pathways by a sequence of chemical reactions and physiological pathways, which are a key factor for the making of solidified sand that is close to natural beachrock, and could be effective for various bioengineering applications [12]. Therefore, for successful implementation of the MICP method in the bio-geoengineering field, environmental factors such as temperature, pH, duration of bacterial cultivation, and crystal precipitation tendency need to be assessed.

In the MICP method, most of the earlier studies addressed ureolytic bacteria as a specific foreign microorganism, and there was a major concern about the adverse effects (for example, competition for nutrients with other species, survival capacity in the new environment, dormancy or sudden shock) on the surrounding environment due to the presence of a large number of foreign species [11–14]. Therefore, to overcome this adverse effect, this study aimed to reduce the impact on ecosystems by isolating urea degradation bacteria from local marine sand (Greece), targeting for coastal erosion protection in that specific area (Greece), and later on we will focus on other areas where coastal erosion is a major problem. Isolated native strains were evaluated for urease activity as whole-cell, supernatant and cell pellets and subsequently different environmental parameters were investigated. On the basis of the results, the sand solidification test (syringe) was also performed, and the degree of solidification was quantitatively evaluated by the needle penetration test. Furthermore, for the practical use of this technology, we proposed a novel approach that is sustainable and eco-friendly, one we aim to apply in future for coastal erosion protection in the targeted area.

Moreover, most of the research on MICP has been limited to land ureolytic bacteria [15–18], focusing on the urea hydrolysis mechanism and on the efficacy of calcite production, and little consideration has been drawn for marine ureolytic bacteria and their subsequent application [5,18–20]. However, some of the previous studies and scope of these studies are mentioned in Table 1. Nonetheless, in the present study, we selected marine ureolytic bacteria isolated from a coastal region in Greece, which has the most comprehensive coastlines among the Mediterranean countries and coastal erosion in this region is severe. In order to protect this long coastal shoreline from erosion, the application of the proposed MICP method was suggested towards the creation of artificial beachrock using native ureolytic bacteria because of its long-term sustainability, considering the local coastal climatic environment. Previously, it was revealed that the formation mechanism of beachrocks is greatly influenced by marine ureolytic bacteria, which show a great affinity for $CaCO_3$ precipitation and can be sustained for a long time, which is a key factor for the MICP method. Hence, we isolated three ureolytic bacterial species (16s rDNA analysis) from the Greek coastal area: *Micrococcus* sp., denoted as G1; *Pseudoalteromonas* sp., denoted as G2; and *Virgibacillus* sp., denoted as G3. We selected native ureolytic bacteria (from Greece) due to their sustainability (considering the temperature, pH, etc.) and high urease activity, which is very important for the making of artificial beachrock by the MICP method.

**Table 1.** Microorganism studied for microbially induced carbonate precipitation (MICP).

| Ureolytic Bacteria | Type | References |
|---|---|---|
| *Sporosarcina pasteurii* | Land | [1,10,18] |
| *Bacillus cohnii* | Land | [14] |
| *Bacillus subtilis* | Land | [12] |
| *Micrococcus* sp. | Marine | This study |
| *Pseudoalteromonas* sp. | Marine | This study |
| *Virgibacillus* sp. | Marine | This study |

The main objectives of this study were to find out an appropriate or feasible condition and method that would be very useful in making artificial beachrocks for coastal erosion protection in Greece and in Mediterranean countries, along with using geotextile tube technology by the MICP method in an inexpensive, eco-friendly, and sustainable way. In future, we aim to apply our findings using our proposed method in the field scale through considering durability study, rainfall effects, low-cost reagent (preferable waste material), and so on. Moreover, the information of this study could play an important source of information for commercial applications in protecting against coastal erosion and for other bio-engineering applications.

## 2. Methodology

### 2.1. Isolation and Bacteria Cell Culture

The ureolytic bacterial (G1, G2, and G3) species were isolated from the coastal area of Porto Rafti and Loutraki in Greece (Figure 1). The soil was sampled from each peripheral beachrock in Greece. After screening, the bacterial species were identified by 16s rDNA gene analysis. The genome extraction was done using 100 μL of Mighty Prep reagent taken into a 1.5 mL microtube. From the colonies on the plate, the cells were removed with a sterilized platinum loop, suspended in the microtube, heated at 95 °C for 10 minutes, and then centrifuged at 12000 rpm for 2 minutes. The supernatant of the suspension centrifuged was transferred to another tube and used as a template for PCR (polymerase chain reaction).

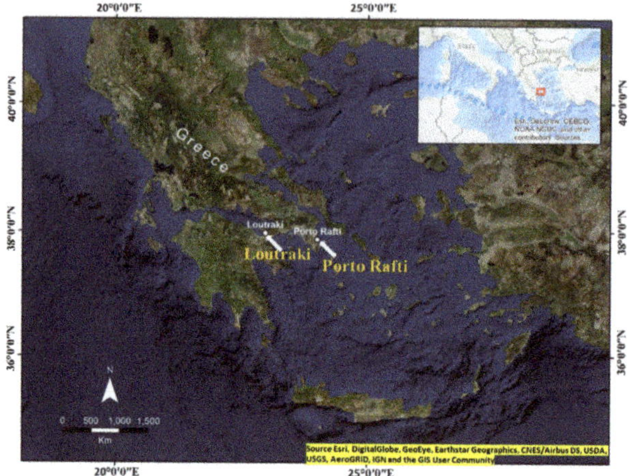

**Figure 1.** Map of Greece showing the sample collection site (Porto Rafti: N 37° 52′ 57.1′′, E 24° 00′ 51.1′′ and Loutraki: N 37° 57′ 02.3′′, E 22° 57′ 37.7′′).

After successful PCR amplification, DNA extraction was carried out by an extraction kit (Nippon Genetics Co. Ltd.). The six types of primers used in this experiment were 9F, 515F, 1099F, 1541R, 1115R, and 536R. Finally, for the identification of the isolated bacterial species 16s rDNA, base sequence analysis (about 1500 bp) was performed. Finally, BLAST (Basic Local Alignment Search Tool) analysis was performed by the database DB-BA12.0 (Techno Suruga Laboratory, Japan). ZoBell2216 culture solution (hi-polypeptone 5.0 g/L, yeast extract 1.0 g/L, and $FePO_4$ 0.1 g/L mixed with seawater (artificial), maintained at pH 7.6–7.8) was used as the main culture medium for the growth of the species. The bacterial cells were pre-cultured (5 mL) in the ZoBell2216 solution at 30°C for 24 h by shaking at 160 rpm. The preculture (1 mL) was inoculated into 100 mL of fresh ZoBell2216 medium and incubated at 30°C with shaking at 160 rpm. Subsequently, it was continuously cultivated for 10 days. During the cultivation, the bacterial cell growth was determined ($OD_{600}$) by a UV-VIS spectrophotometer (V-730, JASCO Corporation, Tokyo, Japan).

### 2.2. Assessment of Urease Activity

Urease activity of the cultured bacterial cells was assessed at different pH and temperature levels following the indophenol method [21,22] where ureolysis occurred directly by the bacterial cell in a solution. The collected sample solution (1 mL) from the cell culture was added to 100 mL of the stratum solution containing 0.3 M urea and 0.1 M EDTA (Basic Local Alignment Search Tool) buffer with phenol-nitroprusside. The reaction composition was stirred using a water bath and kept at a steady temperature and pH. At the sampling point, 10 mL solution was sampled and 0.1 mL of 10 M NaOCl was added to the solution to prevent quick enzymatic hydrolysis. The composed solution was incubated subsequently for 10 min at 50–60 °C. The urease activity test was carried out to evaluate the effect of the reaction solution conditions, by 48 h cultured cells at distinctive temperatures ranging from 10 to 60 °C, and the pH was 6, 7, and 8 (adding EDTA buffer solutions at 30 °C). The urease activity of the discharged urease from the culture supernatants was also investigated. Cultured bacteria cell culture (1.2 mL) was centrifuged by 10,000 rpm for 5 min at 10 °C, and then the supernatant (1 mL) was used to examine urease activity as per the following method. Urease activity of the sampled supernatant would correspond to the discharge of urease. The urase activity of bacterial cell pellets and their relationship to bacterial cell growth was also evaluated.

To remove the supernatant, a 40 mL sample was collected from the bacterial cell culture tube and centrifuged by 10,000 rpm, for 5 min at 10 °C. To get a solution from the resuspended bacterial cell, the cell pellet was washed by 0.1 M Tris and HCl buffer for repeated times. The urease activity assessment was conducted considering different $OD_{600}$ values. The urease activity of the cell lysate, soluble and insoluble portion of the cell lysate, and finally the whole bacterial cell was also investigated.

After 48 h culture of the cell, 10 mL sample solution was collected and centrifuged by following the same procedure as mentioned earlier. The sample solution was resuspended to remove the culture supernatant with 500 mL sonication buffer (0.1 M NaCl, 20 mM Tris/HCl, and 1 mM EDTA) and was named as the whole-cell. The soluble cell lysate was prepared using the disrupted bacterial cell using ultrasonication for 5 min with 130 W, and time interval was 1 min. Remaining bacterial cell-debris was also collected by centrifugation for 20 min at 4 °C with 8000 rpm, which was insoluble to the solution. The urease activity assessment was also conducted for both soluble and insoluble cases, followed by the indophenol method.

### 2.3. Sodium Dodecyl Sulfate-Polyacrylamide Gel Electrophoresis (SDS-PAGE) Analysis

SDS-PAGE (sodium dodecyl sulfate-polyacrylamide gel electrophoresis) was conducted using 10 mL bacterial cell culture solution. Using the ultrasonication method (5 min and 130 W) the cultured bacterial cell was broken and the collected cell was resuspended with 500 mL of buffer solution (sonication). The collected solution (soluble) was moved to a tube for SDS-PAGE analysis after centrifugation for 20 min at 4 °C with 8000 rpm. In the meantime, the precipitate segment of cell lysate was dissolved with denaturation buffer (6 M urea with 50 mM Tris/HCl). The collected insoluble solution was also assessed by SDS-PAGE method. Finally, after electrophoresis, the formed gel was washed with CBB solution (Coomassie Brilliant Blue) by Nacalai Tesque, Inc., Kyoto, Japan.

### 2.4. Microbial $CaCO_3$ Precipitation Test

The selected strains (G1, G2, and G3) were cultured with a ZoBell2216 solution (1.0 g/L of yeast extract, 5.0 g/L of hi-polypeptone, and 0.1 g/L of $FePO_4$) mixed with artificial seawater (Table 2) with a controlled pH of 7.6–7.8, until the maximum cell growth was observed. The cell growth ($OD_{600}$) was measured by a UV-VIS spectrophotometer (V-730, JASCO Corporation, Tokyo, Japan). In the meantime, 0.5 M $CaCl_2$ and 0.5 M urea were put in a test tube up to 10 mL. The bacterial culture solution was centrifuged (12,000 rpm) for 10 min. The supernatant and cell pellets were divided and the bacterial cell concentration ($OD_{600}$) was adjusted using distilled water. The adjusted bacterial cell concentration was then added to the test tube and put in a shaker at 30 °C for 48 h. Finally, the precipitation (white crystals) was observed and the precipitated crystals (centrifuged at 12,000 rpm for 10 min) was separated from the culture solution by a filter paper. Precipitated crystal tubes were kept in an oven drier for 24 h at 100 °C and then the final weight was taken to calculate the total crystal precipitation amount.

**Table 2.** Composition of artificial seawater.

| Reagent | Concentration (g/L) |
|---|---|
| $MgCl_2 \cdot 6H_2O$ | 222.23 |
| $CaCl_2 \cdot 2H_2O$ | 30.7 |
| $SrCl_2 \cdot 6H_2O$ | 0.85 |
| KCl | 13.90 |
| KBr | 2.00 |
| $H_3BO_3$ | 0.54 |
| NaCl | 490.68 |
| $Na_2SO_4$ | 81.88 |

### 2.5. Sand Solidification Test

In order to check the ability of sand solidification of the isolated strains, a small-scale sand syringe solidification test was conducted. First, the isolated strains were grown in ZoBell2216E culture medium and shaken continuously at 30 °C for 2 days.

Then, 40 g of dried (110 °C for 2 days) Mikawa sand was loaded in a 35 mL syringe (height: 7 cm and diameter: 2.5 cm). Subsequently, 16 mL of the bacterial culture solution (ZoBEll2216E) and 20 mL of the consolidation solution (Table 3) was serially injected into the syringe showing in Figure 2a. The syringe was flushed after 2 h, leaving approximately 2 mL of the solution close to the surface of the sand. The consolidation solution was injected and drained every day for 21 days. The testing conditions are presented in Table 4. The pH values and $Ca^{2+}$ concentrations from the outlet were measured afterwards. The condition of the syringe solidification test and grain size distribution of Mikawa sand is presented in Figure 2b. After 21 days, the UCS of the sample was measured by a needle penetration device (SH-70, Maruto Testing Machine Company, Tokyo, Japan).

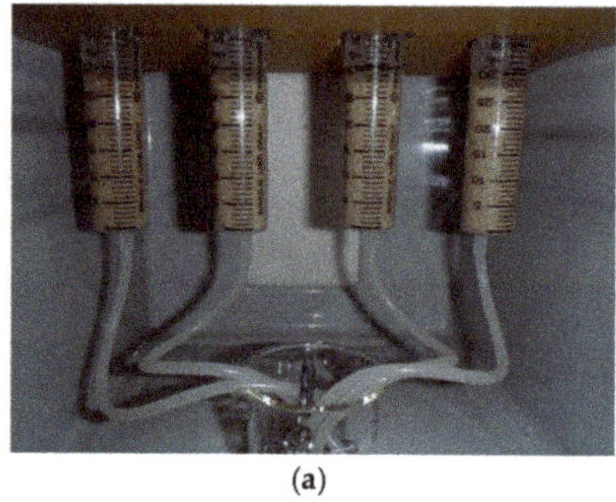
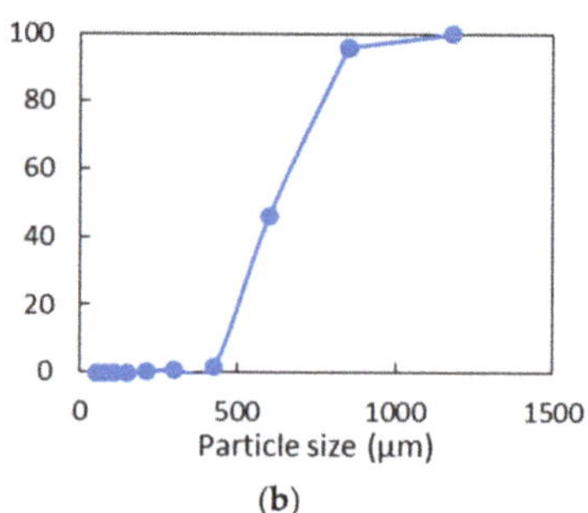


(a)                                  (b)

**Figure 2.** The state of (**a**) sand solidification test (syringe) and (**b**) particle size distribution of Mikawa sand.

**Table 3.** Composition of sand solidification solution.

| Composition | Concentration (g/L) |
| --- | --- |
| $CO(NH_2)_2$ | 30.0 |
| $CaCl_2$ | 55.5 |
| Nutrient broth | 3.0 |
| $NaHCO_3$ | 2.12 |
| $NH_4Cl$ | 10.0 |

**Table 4.** Testing conditions for sand solidification test (syringe).

| Test Period | Incubation Time | Temperature | Solidification Solution Concentration | Nutrient Injection Interval | Solidification Solution Injection Interval |
| --- | --- | --- | --- | --- | --- |
| 21 days | 24 h<br>48 h<br>72 h | 30 °C | 0.5 M | -<br>7 days<br>- | -<br>1 day<br>- |

## 3. Results and Discussion

### 3.1. Cell Concentration (Isolated Species) and Urease Activity

According to the BLAST analysis result, the identified species are shown in Table 5. Urease activity is closely related to the bacterial cell growth and it was revealed that a high bacterial cell concentration enhanced the amount of calcite precipitated by the MICP method. Moreover, it was concluded that the urea hydrolysis rate was directly proportionate to the concentration of bacteria, which is an important factor for the success of the MICP application [23,24]. On the basis of these findings, the generation of urease from the isolated species could be a growth correlated framework that favors for sand solidification. Figure 3a represents the cell growth, whereas Figure 3b,c illustrate the urease activity of the G1, G2, and G3 species as whole-cell, cell pellet, and supernatant, respectively. From Figure 3a, it can be observed that initially the bacterial cell growth increased with time and then became stable and, conversely, the urease activity increased initially and then decreased, making a bell-shaped profile in between 3–6 days for the G1, G2, and G3 species.

**Table 5.** Identified strains from the collected soil sample.

| Code Name | Species |
| --- | --- |
| G1 | *Micrococcus* sp. |
| G2 | *Pseudoalteromonas* sp. |
| G3 | *Virgibacillus* sp. |

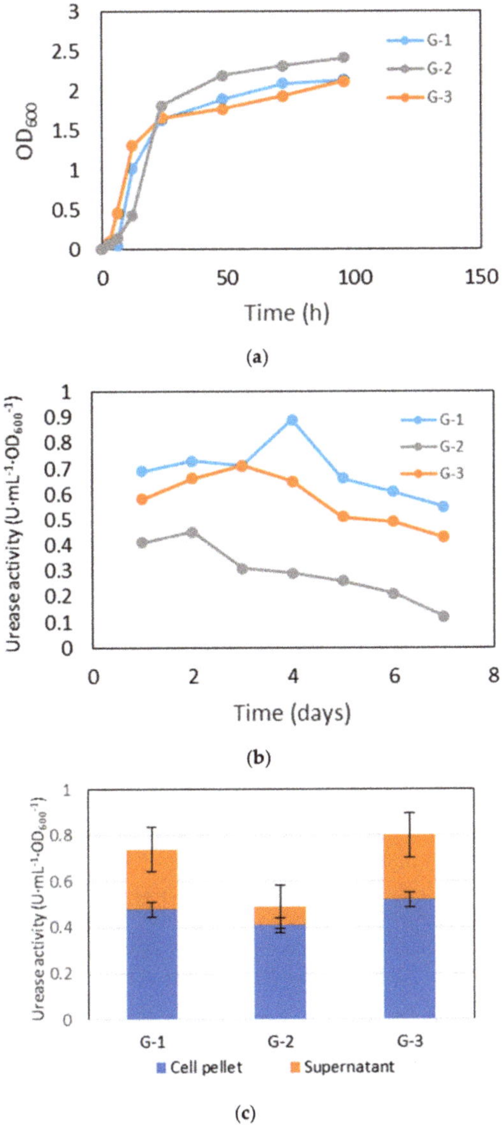

**Figure 3.** The results showing the (**a**) cell growth and (**b**) urease activity of the isolated strains including the (**c**) cell pellets and supernatant.

It was shown that the urease activity of the bacterial whole cell, cell pellet, and supernatant was closely related to the presence of protease in the cell [25]. However, it is well established that urease activity of different bacteria species depends on each species, as well as some environmental factors, which is similar to this study. Earlier studies have also reported that the urease from a few species such as *Pararhodobacter* sp. Appeared to have very high stability with extensive cultivation [26]. However, our results revealed that the G2 species of bacteria showed a dissimilar tendency for urea hydrolysis even under extended cultivation compared to the G1 and G3 species. The possible reasons for varying the urease activity might be osmotic pressure and resuspension of the cell, which might release enzyme into the solution from cells. However, further study needs to investigate the actual reasons.

The findings of this study are consistent with earlier findings and could contribute as useful information for the sand solidification process by MICP, considering the bacteria population increase via centrifuge and/or applying the re-injecting method of bacteria before urease activity becomes null.

### 3.2. Impact of pH to Urease Activity

Figure 4 represents urease activity with different pH conditions at a certain temperature (30 °C) for G1, G2, and G3 species. The results exposed a nearly bell-shaped outline showcasing the maximum activity around pH 7 for all of the three species.

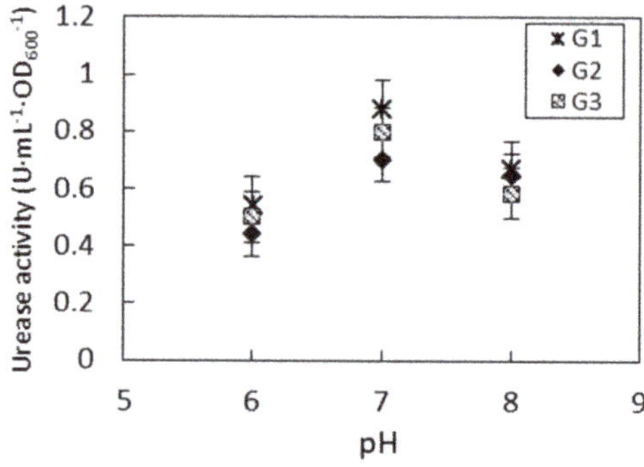


**Figure 4.** Impact of pH for the urease activity of the reaction solution at 30 °C.

Following the same conditions, the G1 species showed the highest urease activity compared to the G2 and G3 species. Previous research showed that the MICP process favored an initial pH environment of 6.5–9.3 (neutral to basic condition). Additionally, it was revealed that, in the case of some species such as *Pararhodobacter* sp., considerable loss of urease activity was observed with acidic (lower than pH 6) and alkaline (higher than pH 9) conditions attributed to denaturation of protein [24]. It was also reported that the urease activity of some ureolytic bacterial species favored lightly alkaline conditions [25–27]. Thus the pH value played an important role for the bacteria transportation and adhesion to obtain homogeneously improved strength of treated soils. Therefore, the results of this study could also be useful information for obtaining a successful sand solidification process for identifying the optimum pH condition for native species.

### 3.3. Impact of Temperature to Urease Activity

Urease activity with various temperature conditions is shown in Figure 5 for the G1, G2, and G3 species. High urease activity was observed at a temperature above 30 °C (G1 species) and then gradually decreased and showed an almost bell-shaped profile because of cell denaturation at a higher temperature, except in the case of G2 and G3 species. The divergent tendency of urease activity (nearly 40–60 °C) was also observed by Fujita et al. [26]. However, there are few data about the urease activity of marine ureolytic bacteria in Greece. Even though the urease activity of G1, G2, and G3 species are lower than that of *Pararhodobacter* sp., the results obtained in this study would be innovative and adventitious information for the sand solidification method through microbial carbonate precipitation because the use of native ureolytic species (isolated from marine area). In future, it may be possible to use G1, G2, and G3 species for sand solidification, which could help to manufacture artificial beachrock following the MICP technique. This result could also be adventitious in Greece for coastal erosion control purposes by increasing bacterial population either by centrifugation and/or urease enzyme extraction by ultrasonication and/or adding bacterial populations several times during the sand solidification process before urease activity completely declined. Similar results were also found for the strains such as *Deleya venusta* and *Pararhodobacter* sp. [26–28]. Previously, it was also found that urease activity influenced to $CaCO_3$ precipitation at temperatures from 10 to 60 °C, which is similar to the findings of this study. However, the effectiveness of environmental factors on urease activity of the negative species with variation temperatures was not examined. Therefore, on the basis of these experimental findings, the G1, G2, and G3 species showed high urease activity in the different range of temperatures that could reflect the importance of isolating the ureolytic bacterial species from the local environment and their subsequent application for sand solidification and soil improvement.

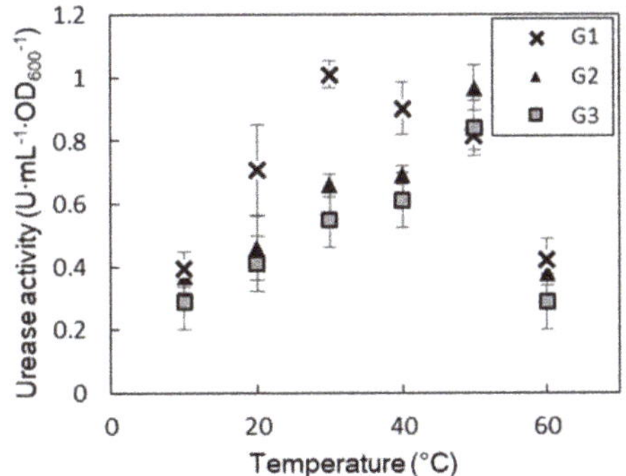


**Figure 5.** Results showing the urease activity of the isolated species at different temperatures.

These results could be very useful for the sand solidification by the MICP method in Mediterranean coastal regions considering the local environmental conditions because the isolated ureolytic bacterial species are well adapted to the local environmental conditions. These results could also be very advantageous if we want to focus on soil or ground improvement techniques by bio-stimulation or bio-augmentation method [29], and we must consider the native ureolytic bacterial species which are well adapted with the local environment and could ensure long term sustainability.

### 3.4. SDS-PAGE Analysis

The SDS-PAGE analysis was performed to confirm the expression proteins for both soluble and insoluble fractions of the strains cell lysate from the isolated native G1, G2, and G3 species, which is vital in detecting the location of the reaction in a mixture of sand/soil particles. Figure 6 represents the SDS-PAGE results for the soluble and insoluble portion of the bacterial cell suspensions.

It has been studied that ureases from bacterial cell are composed of two small subunits (8–12 kDa) with one subunit being larger [30], and the results of SDS-PAGE analysis exhibited several bands of nearly 15 kDa and 60 kDa [26,29]. These results indicated that the bacterial urease could be expressed as either insoluble protein or as a soluble membrane protein, which also corresponds to the results [26,31].

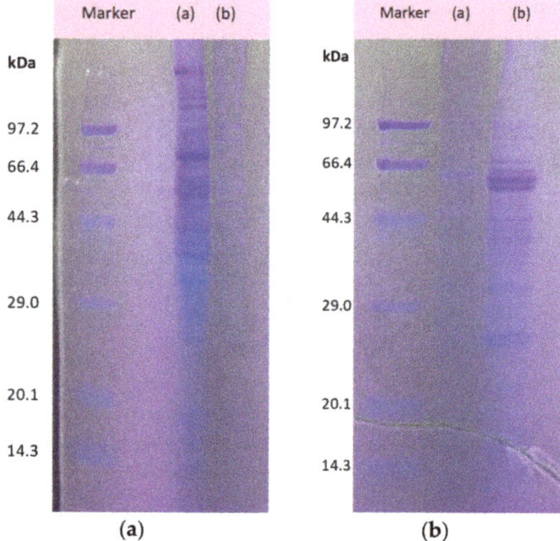

**Figure 6.** Confirmation of protein (enzyme) expression of the (**a**) soluble and (**b**) insoluble portion of bacterial cell lysate of the isolated species.

These results showed that the variation of urease activity depended on cell lysate, cell membrane, and whole-cell by focusing on the urea hydrolysis by native bacterial strain.

Thus, on the basis of these results, urease could exist as a soluble protein or likely as a soluble membrane protein. We found that the protein was not released into the culture solution but was separated from the bacterial cell membrane into the solution during ultrasonication because the crystal precipitation was greatly influenced by the bacterial cells. Therefore, the findings in this study support in figuring out the urease accumulation characteristic and extraction of the enzyme by sonication, which could be effective for the efficacy of the soil treatment/ground improvement by MICP and/or enzymatic process in practical field application for coastal erosion protection, especially in the coastal area of Greece.

*3.5. Microbial CaCO$_3$ Precipitation Test*

Earlier studies showed that it is possible to control the strength of the treated sand by regulating the amount and/or volume of precipitated minerals. It has also been reported that various environmental factors, including temperature, pH, bacterial size, and cell concentration, control the amount of calcite precipitation during the MICP process [31–33]. Therefore, it is very important to investigate the trend in CaCO$_3$ precipitation for individual bacterial species because the CaCO$_3$ acts as the main binder material between the substrate particles for soil improvement during the MICP process. The CaCO$_3$ precipitation trend for the G1, G2, and G3 species is shown in Figure 7a. From the figure, it is clear that CaCO$_3$ precipitation is closely related to the OD$_{600}$ values, which supports previous studies [34]. From our studies, was also observed that higher OD$_{600}$ values influenced the CaCO$_3$ precipitation (Figure 7b) amount. Under the same conditions, the G1 and G3 species had the advanced ability of precipitating a higher amount of CaCO$_3$ compared to the G2, leading to a higher possibility of sand solidification. At the same time, the rate of the precipitated CaCO$_3$ decreased and tended to be constant with a further increase of the bacterial cell concentration. However, this study suggests that the variation in the CaCO$_3$ precipitation trend was related to the bacterial population of individual species, which supports earlier studies.

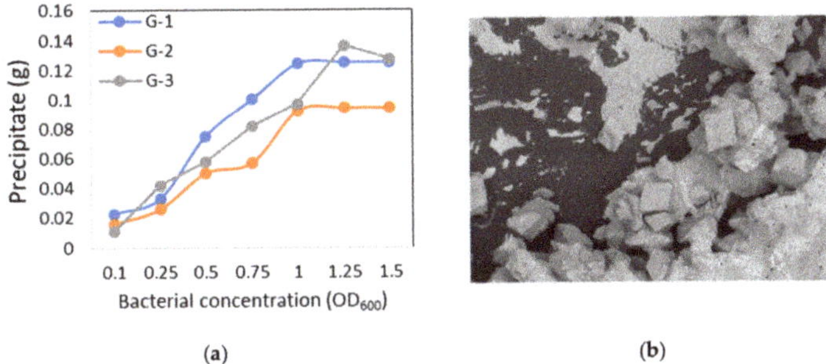

(a)                                  (b)

**Figure 7.** (**a**) CaCO$_3$ precipitation trend of the isolated strains and (**b**) SEM images of the precipitated CaCO$_3$ of G1 species.

### 3.6. Potentiality for Sand Solidification and Design for the Application

The results of the temporal variation of Ca$^{2+}$ and the pH effluent is presented in Figure 8a,b, respectively. From the figure, it is shown that the Ca$^{2+}$ discharge liquid was seen to continue to rise from around 4 days from bacterial implantation and then decreased with time (Figure 8a). The Ca$^{2+}$ concentration increased, indicating that the hydrolysis effect of the urea-degrading bacteria in the syringe no longer worked, and it was presumed that the injection of regular cells was required. The pH tended to increase and then decrease with time (Figure 8b) which indicated the variation of urea hydrolysis inside the syringe.

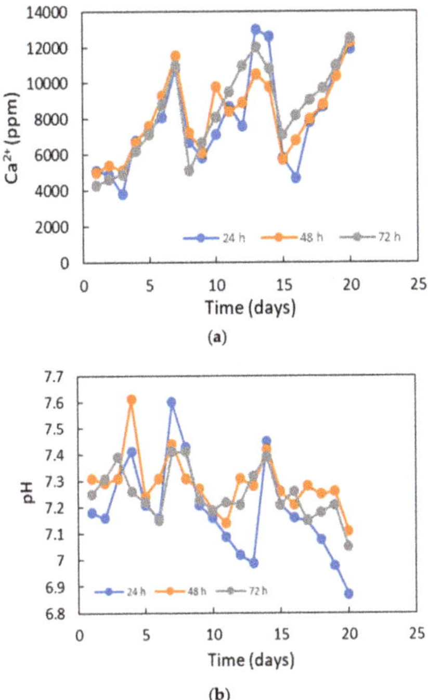

(a)

(b)

**Figure 8.** (**a**) Temporal variation of Ca$^{2+}$ and (**b**) pH in the effluent from solidification solution in days.

The outcomes of the needle penetration test are summarized individually and shown in Figure 9. From the figure, it was observed that maximum USC around 1.8 MPa was obtained at the middle portion for the G3, G1, and G2 species. The top portion was moderately solidified (UCS around

1.3 MPa) and the bottom portion was solidified bearing UCS around 1.4 MPa (for G3). This was because maximum urea hydrolysis possibly occurred at the middle portion and precipitated crystal might have had more contact point to bind sand particles at this stage [35]. Another reason could be the variation of aerobic and anaerobic conditions in between the bottom portion and beneath the portion of the sample that might be responsible in decreasing the urease activity and/or crystal precipitation. Therefore, further investigation is required to improve the solidification condition.

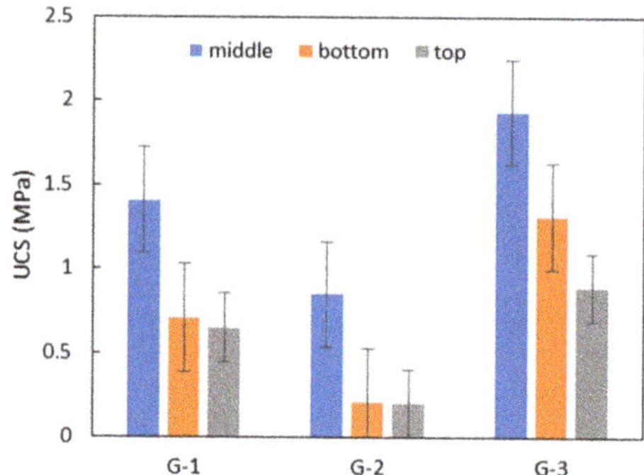

**Figure 9.** Comparison of estimated unconfined compressive strength (UCS) in each strain.

It is well known that the traditional geotubes are widely utilized for flood control, to prevent waterlogging conditions, and for coastal erosion protection. However, all of these conventional methods are not sustainable because of the various of UV rays that come from sunlight damage the geotube after a certain period of time, and protection of soil might be hampered and thus, consequently, the re-installing cost for geotube will be increased [36]. Nonetheless, the re-installing cost could be minimized by using the MICP method and the geotube following our proposed novel approach (Figure 10). By implementing this novel approach, it would be possible to strengthen the coastal structure because the geotube and solidified sand MICP will protect against the UV rays, helping it to remain for a longer period and being more sustainable than the conventional method. In our proposed implementation method, the total cost will be much lower compared to traditional methods, and local ureolytic bacteria will be used, which is more eco-friendly and will ensure long term sustainability of the coastal structures. This method can also be useful in a wider range of civil engineering implementations including paving, construction, and drainage.

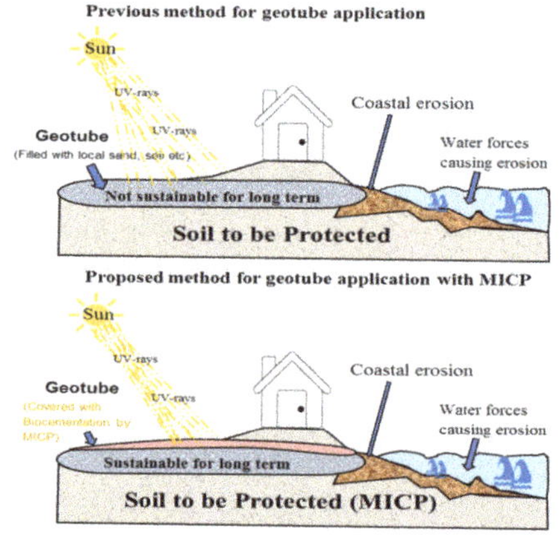


**Figure 10.** Proposed coastal erosion protection method.

## 4. Conclusions

From our experimental results, we demonstrated that the G3 species showed a potentially advanced growth rate and urease activity when compared to the G1 and G2 species. Nonetheless, the environmental parameters also had a remarkable consequence on the bacterial urease activity. Moreover, from the SDS-PAGE analysis, it was revealed that urease could be released as a soluble protein or conceivably as a membrane protein (soluble) in/on the bacterial cell, but the exact location of urease activity could not be detected. We also concluded that urease activity for G1, G2, and G3 species relied on the environmental parameters. Additionally, urease was not secreted into the culture solution but accumulated in and/or on the cell, and the urease activity was also observed in resuspended cells. Indeed, the unconfined compressive strength (UCS) of our solidified sand was lower than when compared to conventional structure. Moreover, its capability could play an important role as a new eco-friendly design for coastal erosion protection in combination with geotube in a more sustainable way. This research showed a positive result leading to the enhanced possibility of application in coastal erosion protection by assimilation of knowledge from microbial mechanism and design applications considering the engineering viewpoint. Future investigations will be focused on studying strength improvements, enhanced durability, and feasibility at a small or medium scale, which are essential for implementation as an ideal substantial design. Moreover, these results could be applied towards the creation of artificial beachrocks by the MICP method for coastal protection in Mediterranean countries, as well as for bio-mediated soil improvement.

**Author Contributions:** All the co-authors contributed equally to designing the experiments, analyzing the data, writing the manuscript, and completing the revisions and editing. All the co-authors have read and agreed with the submitted version of the manuscript.

**Funding:** This work was partly supported by JSPS KAKENHI, grant number JP16H04404.

**Acknowledgments:** The authors deeply express their acknowledgments to all the laboratory members of Biotechnology for Resource Engineering, Hokkaido University, for their generous support, and the collaborating Faculty of Geology and Geoenvironment, National and Kapodistrian University of Athens, Athens, Greece, and all other advisers and reviewers.

**Conflicts of Interest:** The authors acknowledge no conflict of interest.

## References

1. Salifu, E.; MacLachlan, E.; Iyer, K.R.; Knapp, C.W.; Tarantino, A. Application of microbially induced calcite precipitation in erosion mitigation and stabilisation of sandy soil foreshore slopes: A preliminary investigation. *Eng. Geol.* **2016**, *201*, 96–105. [CrossRef]
2. Evelpidou, N.; Vassilopoulos, A.; Leonidopoulou, D.; Poulos, S. AN INVESTIGATION OF THE COASTAL EROSION CAUSES IN SAMOS ISLAND, EASTERN AEGEAN SEA. *Tájökológiai Lapok* **2008**, *6*, 295–310.
3. Mujah, D.; Shahin, M.A.; Cheng, L. State-of-the-Art Review of Biocementation by Microbially Induced Calcite Precipitation (MICP) for Soil Stabilization. *Geomicrobiol. J.* **2017**, *34*, 524–537. [CrossRef]
4. Danjo, T.; Kawasaki, S. Formation mechanisms of beachrocks in Okinawa and Ishikawa, Japan, with a Focus on Cements. *Mater. Trans.* **2014**, *55*, 493–500. [CrossRef]
5. Danjo, T.; Kawasaki, S. Microbially Induced Sand Cementation Method Using Pararhodobacter sp. Strain SO1, Inspired by Beachrock Formation Mechanism. *Mater. Trans.* **2016**, *57*, 428–437. [CrossRef]
6. Sheehan, C.; Harrington, J. An environmental and economic analysis for geotube coastal structures retaining dredge material. *Resour. Conserv. Recycl.* **2012**, *61*, 91–102. [CrossRef]
7. Ashis, M. Application of geotextiles in Coastal Protection and Coastal Engineering Works: An overview. *Int. Res. J. Environ. Sci.* **2015**, *4*, 96–103.
8. Lee, S.C.; Hashim, R.; Motamedi, S.; Song, K.-I. Utilization of geotextile tube for sandy and muddy coastal management: A review. *Sci. World J.* **2014**, *2014*, 494020. [CrossRef]
9. Kubo, R.; Kawasaki, S.; Suzuki, K.; Yamaguchi, S.; Hata, T. Geological Exploration of Beachrock through Geophysical Surveying on Yagaji Island, Okinawa, Japan. *Mater. Trans.* **2013**, *55*, 342–350. [CrossRef]

10. Neumeier, U. Experimental modelling of beachrock cementation under microbial influence. *Sediment. Geol.* **1999**, *126*, 35–46. [CrossRef]

11. Whiffin, V.S.; van Paassen, L.A.; Harkes, M.P. Microbial carbonate precipitation as a soil improvement technique. *Geomicrobiol. J.* **2007**, *24*, 417–423. [CrossRef]

12. Sarayu, K.; Iyer, N.R.; Murthy, A.R. Exploration on the biotechnological aspect of the ureolytic bacteria for the production of the cementitious materials—A review. *Appl. Biochem. Biotechnol.* **2014**, *172*, 2308–2323. [CrossRef] [PubMed]

13. Tiano, P.; Biagiotti, L.; Mastromei, G. Bacterial bio-mediated calcite precipitation for monumental stones conservation: Methods of evaluation. *J. Microbiol. Methods* **1999**, *36*, 139–145. [CrossRef]

14. Achal, V.; Mukherjee, A.; Basu, P.C.; Reddy, M.S. Strain improvement of Sporosarcina pasteurii for enhanced urease and calcite production. *J. Ind. Microbiol. Biotechnol.* **2009**, *36*, 981–988. [CrossRef]

15. Jonkers, H.M.; Thijssen, A.; Muyzer, G.; Copuroglu, O.; Schlangen, E. Application of bacteria as self-healing agent for the development of sustainable concrete. *Ecol. Eng.* **2010**, *36*, 230–235. [CrossRef]

16. Van Paassen, L.A.; Ghose, R.; van der Linden, T.J.M.; van der Star, W.R.L. Quantifying Biomediated Ground Improvement by Ureolysis: Large-Scale Biogrout Experiment. *J. Geotech. Geoenviron. Eng.* **2010**, *136*, 1721–1728. [CrossRef]

17. Harkes, M.P.; van Paassen, L.A.; Booster, J.L.; Whiffin, V.S.; van Loosdrecht, M.C.M. Fixation and distribution of bacterial activity in sand to induce carbonate precipitation for ground reinforcement. *Ecol. Eng.* **2010**, *36*, 112–117. [CrossRef]

18. Omoregie, A.I.; Ngu, L.H.; Ong, D.E.L.; Nissom, P.M. Low-cost cultivation of Sporosarcina pasteurii strain in food-grade yeast extract medium for microbially induced carbonate precipitation (MICP) application. *Biocatal. Agric. Biotechnol.* **2019**, *17*, 247–255. [CrossRef]

19. Lee, C.; Lee, H.; Kim, O.B. Biocement fabrication and design application for a sustainable urban area. *Sustainability* **2018**, *10*, 4079. [CrossRef]

20. Ge, P.; Nayanthara, N.; Buddhika, A.; Dassanayake, N.; Nakashima, K.; Kawasaki, S.; Lanka, S.; Author, C. Biocementation of Sri Lankan Beach Sand Using Locally Isolated Bacteria: A Baseline Study on The. *Int. J. GEOMATE* **2019**, *17*, 55–62.

21. Wei, S.; Cui, H.; Jiang, Z.; Liu, H.; He, H.; Fang, N. Biomineralization processes of calcite induced by bacteria isolated from marine sediments. *Braz. J. Microbiol.* **2015**, *46*, 455–464. [CrossRef] [PubMed]

22. Novitsky, J.A. Calcium carbonate precipitation by marine bacteria. *Geomicrobiol. J.* **1981**, *2*, 375–388. [CrossRef]

23. Natarajan, K.R. Kinetic Study of the Enzyme Urease from Dolichos biflorus. *J. Chem. Educ.* **2009**, *72*, 556. [CrossRef]

24. Mortensen, B.M.; Haber, M.J.; Dejong, J.T.; Caslake, L.F.; Nelson, D.C. Effects of environmental factors on microbial induced calcium carbonate precipitation. *J. Appl. Microbiol.* **2011**, *111*, 338–349. [CrossRef] [PubMed]

25. Ferris, F.G.; Phoenix, V.; Fujita, Y.; Smith, R.W. Kinetics of calcite precipitation induced by ureolytic bacteria at 10 to 20°C in artificial groundwater. *Geochim. Cosmochim. Acta* **2004**, *68*, 1701–1710. [CrossRef]

26. Fujita, M.; Nakashima, K.; Achal, V.; Kawasaki, S. Whole-cell evaluation of urease activity of Pararhodobacter sp. isolated from peripheral beachrock. *Biochem. Eng. J.* **2017**, *124*, 1–5. [CrossRef]

27. Zhu, T.; Dittrich, M.; Zhang, W.; Bhubalan, K.; Dittrich, M.; Zhu, T. Carbonate Precipitation through Microbial Activities in Natural Environment, and Their Potential in Biotechnology: A Review. *Front. Bioeng. Biotechnol.* **2016**, *4*, 4. [CrossRef]

28. Jahns, T. Urea uptake by the marine bacterium Deleya venusta HG1. *J. Gen. Microbiol.* **2009**, *138*, 1815–1820. [CrossRef]

29. Adams, G.O.; Fufeyin, P.T.; Okoro, S.E.; Ehinomen, I. Bioremediation, Biostimulation and Bioaugmentation: A Review. *Int. J. Environ. Bioremediat. Biodegrad.* **2015**, *3*, 28–39.

30. Clemens, D.L.; Lee, B.Y.; Horwitz, M.A. Purification, characterization, and genetic analysis of Mycobacterium tuberculosis urease, a potentially critical determinant of host-pathogen interaction. *J. Bacteriol.* **1995**, *177*, 5644–5652. [CrossRef]

31. Silva-Castro, G.A.; Uad, I.; Rivadeneyra, A.; Vilchez, J.I.; Martin-Ramos, D.; González-López, J.; Rivadeneyra, M.A. Carbonate Precipitation of Bacterial Strains Isolated from Sediments and Seawater: Formation Mechanisms. *Geomicrobiol. J.* **2013**, *30*, 840–850. [CrossRef]

32. Deng, W.; Wang, Y. Investigating the Factors Affecting the Properties of Coral Sand Treated with Microbially Induced Calcite Precipitation. *Adv. Civ. Eng.* **2018**, *2018*, 9590653. [CrossRef]
33. Kadhim, F.J.; Zheng, J.J. Review of the Factors That Influence on the Microbial Induced Calcite Precipitation. *Civ. Environ. Res.* **2016**, *8*, 69–76.
34. Keykha, H.A.; Asadi, A.; Zareian, M. Environmental Factors Affecting the Compressive Strength of Microbiologically Induced Calcite Precipitation-Treated Soil. *Geomicrobiol. J.* **2017**, *34*, 889–894. [CrossRef]
35. Li, L.; Zhao, Q.; Zhang, H.; Amini, F.; Li, C. A Full Contact Flexible Mold for Preparing Samples Based on Microbial-Induced Calcite Precipitation Technology. *Geotech. Test. J.* **2014**, *37*, 917–921.
36. Al Imran, M. Combination Technology of Geotextile Tube and Artificial Beachrock for Coastal Protection. *Int. J. GEOMATE* **2017**, *13*, 67–72. [CrossRef]

*Article*

# Video Monitoring of Shoreline Positions in Hujeong Beach, Korea

**Yeon S. Chang [1], Jae-Youll Jin [2], Weon Mu Jeong [3], Chang Hwan Kim [4] and Jong-Dae Do [2,***

[1]  Coastal Development and Ocean Energy Research Center, Korea Institute of Ocean Science and Technology, Busan 49111, Korea; yeonschang@kiost.ac.kr
[2]  Coastal Morphodynamics Team, Korea Institute of Ocean Science and Technology, Busan 49111, Korea; jyjin@kiost.ac.kr
[3]  Maritime ICT R & D Center, Korea Institute of Ocean Science and Technology, Busan 49111, Korea; wmjeong@kiost.ac.kr
[4]  Dokdo Research Center, Korea Institute of Ocean Science and Technology, Busan 49111, Korea; kimch@kiost.ac.kr
*  Correspondence: jddo@kiost.ac.kr; Tel.: +82-54-780-5305

Received: 1 November 2019; Accepted: 18 November 2019; Published: 20 November 2019

**Featured Application: The video monitoring system and mobile light detecting and ranging (LiDAR) used in this study can be applied for erosion control in beaches with various bedforms and coastal structures.**

**Abstract:** Shoreline processes observed by a video monitoring system were investigated under different wave conditions. A 30 m-high tower equipped with video cameras was constructed in Hujeong Beach, South Korea, where coastal erosion was suspected to occur. Two-year shoreline data since December 2016 showed that beach area, $A_b$, has decreased, but periods of rapid increase in $A_b$ were also observed. Shoreline change was closely related to the wave propagation directions and bottom topography. $A_b$ increased when waves approached the shore obliquely, whereas it decreased when they approached in a normal direction. The shoreline became undulated when $A_b$ increased, while it became flatter when $A_b$ decreased. The undulation process was influenced by nearshore bedforms because the shoreline protruded in the lee area where underwater rocks or nearshore sandbars actively developed, with a sheltering effect on waves. Specifically, the locations of shoreline accretion corresponded to the locations where the sandbar horns (location where a crescentic sandbar protrudes toward the shore) developed, confirming the out-of-phase coupling between sandbars and shoreline. When waves with higher energy approached normal to the shore, the sheltering effect of sandbars and underwater rocks became weaker and offshore sediment transport occurred uniformly along the coast, resulting in flatter shorelines.

**Keywords:** video monitoring; coastal erosion; shoreline change; nearshore sandbars

---

## 1. Introduction

Coastal erosion is a serious problem that may cause economic and environmental damage to the coastal communities. For proper protection of coastal environments, a systematic management of the coastal zones must be implemented by the government together with local communities. Such management strategies should be built based on scientific information, which provides proper tools to understand the status of the coasts, and accurate monitoring of the on-going coastal processes, which is required to produce long-term data.

Since the first technique using remotely sensed video images was introduced to measure the morphology and scale of sand bars [1], video monitoring techniques have been widely applied to detect shoreline positions [2,3] as a tool for coastal zone management [4]. For example, 40 video monitoring systems have been employed in the beaches along the coasts of South Korea to calculate the seasonal and annual variation of beach areas, and these data have been used to develop coastal management plans.

Along with their applications for coastal management, video systems have also been used for scientific applications to measure nearshore hydrodynamic processes such as longshore currents [5] and wave parameters [6,7]. In addition, they have been applied successfully in engineering practices such as evaluating the performance of beach nourishment [8]. Video systems are also advantageous to monitor short-term beach processes. During extreme storm events, the shoreline could be rapidly changed in a few days, and the erosion and recovery process could be analyzed using video cameras [9]. One of the greatest benefits of video monitoring over point monitoring is its wider spatial coverage achieved by processing obliquely captured image data [10]. Specifically, it provides significant advantages in monitoring complex shorelines where nearshore processes do not occur homogeneously, and thus, the shoreline change pattern becomes irregular due to the existence of rocky seabeds or nearshore sandbars.

In this study, we analyze the shoreline variations in Hujeong Beach, South Korea, where a video monitoring system has been operating since December 2016. Hujeong Beach is characterized by active development of nearshore sandbars and as a result, its shoreline has a complex pattern with significant variations over time. In addition, there are parts of the seabed that are covered by rocks, although most parts are covered by sand, which makes the shoreline pattern more complicated. The goal of this study is, therefore, to examine the shoreline variations in Hujeong Beach that have been specifically affected by the existence of nearshore sandbars and underwater rocks in relation to the wave conditions. Although the video monitoring system itself is similar to those employed in previous studies, we believe the results of the present work are valuable as the shoreline variation data shows characteristic patterns according to the combined effects of bedforms and waves. Therefore, we used the wave data that were continuously measured during the two years of the experimental period at approximately 1 km offshore from the coast, which increased the reliability of the data in correlation with the shorelines. In addition, the shoreline positions estimated by the video cameras may contain errors when averaging the images to remove noise; thus, validation of the video data is required. For validation, we additionally employed a ship-mounted mobile light detecting and ranging (LiDAR) instrument to measure the beach topography outside the water.

## 2. Video Monitoring System

The video monitoring system was constructed in the middle of Hujeong Beach, which is on the east coast of South Korea (Figure 1a). In the northern end of the beach, an intake breakwater is located to protect the intake system of the Hanul nuclear power plant (NPP), as shown in Figure 1b [11]. This breakwater does, however, block the sediments that are discharged from Bugu River to be transported south to Hujeong Beach. Therefore, the beach is vulnerable to erosion because the input amount of sediments has decreased due to the breakwater whereas they are still lost to the southeast of the beach through Jukbyeon Port. To monitor this process, the East Sea Research Institute (ESRI) of the Korea Institute of Ocean Science and Technology (KIOST) has operated the video monitoring system since December 2016.

In total, eight video cameras were installed on top of a 30 m-high metal tower. The height of the tower was determined to obtain secure angles for the cameras because berms have occasionally formed along the beach shore causing blind spots. The coverage of the monitoring system is ~2.4 km, with ~1.2 km at either side of the tower. The cameras capture images every second. Figure 2a shows a picture of the video monitoring tower constructed at S1, and a magnified view of the top of the tower where the video cameras were installed is shown in Figure 2b. In Figure 2c–i, snapshots taken by each of the cameras are shown as an example of the image data. To extract the shoreline data more precisely, 600 snapshot images are averaged over 10 min to remove the noise produced by wave breaking [12].

Every hour, only the first 10 min of data were averaged to produce one representative image of the hour. As the images were not visible during the night, only 12–15 image data were available per day to be processed for the estimation of shoreline positions. The image data were transferred to the processing center using an internet network system [13]. Once the 600 snapshot images were averaged, the shoreline area appeared as a white narrow band due to the swash zone. The shorelines were manually extracted from these averaged images by taking a line in the middle of each band. Because errors in this extraction process might increase when the waves were actively breaking, the extracted shoreline data were accepted only when the band widths were narrow under mild wave conditions (wave heights < 0.5m).

**Figure 1.** (a) Location of Hujeong Beach; (b) map of Hujeong Beach with location of the video monitoring system (L1). This figure is taken from [11].

**Figure 2.** (a,b) Photographs of the video monitoring tower installed in Hujeong Beach, December 2016. (c–i) Examples of shoreline image data measured by the video cameras. These images are taken from [12].

The shoreline positions that are obtained by averaging the snapshot images may contain errors when removing the noise by wave breaking, which requires validation of the data. Figure 3 shows a comparison of the shoreline position data estimated by the video monitoring system with beach topography data outside the water. The land topography data were measured by a mobile LiDAR that was mounted on a boat on 12 March 2017. The mobile LiDAR was equipped with a real time kinematic (RTK), global navigation satellite system (GNSS), and inertia measurement unit (IMU). Ship-mounted mobile LiDARs have advantages over the land-based mobile LiDARs as they can survey a larger area relatively fast by exploring along the coastline. The mobile LiDAR has a measurement range of 1 km and a scanning range up to 80° in the vertical and up to 360° in the horizontal, with accuracy of 2 cm. As shown in Figure 3, the shoreline positions estimated by the video monitoring system accurately match the positions measured by the mobile LiDAR. Discrepancies can be found specifically near the areas where the horizontal gradient of the shoreline is relatively large. However, the maximum error of the shoreline position is of approximately 10 m and the resulting error in the estimation of beach area is less than 3%, which validates the application of video monitoring data for this study.

**Figure 3.** Comparison of shoreline positions estimated with the video monitoring system (red solid line) with land topography (green and yellow contours) measured by a ship-mounted mobile LiDAR on 12 March 2017. The seaward boundary of the colored contours determines the shoreline positions.

## 3. Experiment

The image data collected by the video monitoring system have been processed to calculate the area of Hujeong Beach, $A_b$, since December 2016 by positioning the coastline from the averaged images. In Figure 4, the time variation of the calculated beach area is plotted until October 2018. During the period of approximately two years, $A_b$ generally decreased, although there were periods when $A_b$ significantly increased from October 2017 to February 2018. From this time series of $A_b$, we selected four terms (T1, T2, T3, and T4) that showed a specific pattern of beach area variation. T1 was chosen during the period from 9 December 2016 to 18 March 2017, when $A_b$ significantly fluctuated with an increase in the resulting area at the end of the term. In T2 (8 June 2017 to 29 October 2017), $A_b$ steadily decreased, losing a total beach area of ~10%. T3 was selected because $A_b$ increased rapidly during a relatively short period from 29 October 2017 to 23 December 2017. On the contrary, $A_b$ significantly decreased during T4 from 13 March 2018 to 15 August 2018. The start and end times of each period are listed in Table 1, and the measurements of $A_b$ at the corresponding time steps are listed in Table 2.

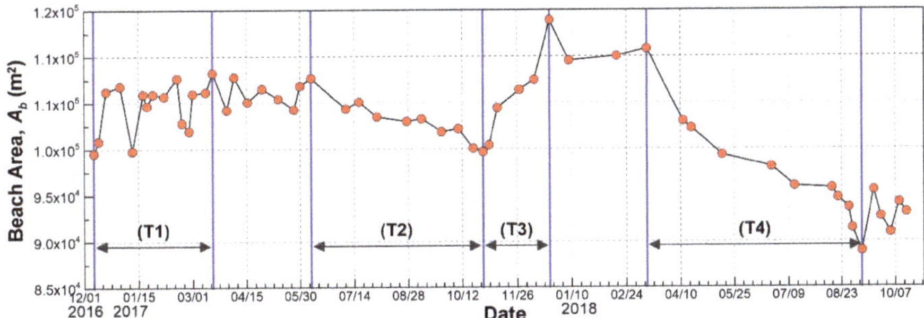

**Figure 4.** Time variation of beach area, $A_b$, from 1 December 2016 to 7 October 2018. T1, T2, T3, and T4, marked by the blue vertical lines, denote the time periods selected for the analysis during which $A_b$ shows a specific variation pattern.

**Table 1.** Start and end dates of the four selected periods.

| Dates | T1 | T2 | T3 | T4 |
|---|---|---|---|---|
| Start | 9 December 2016 | 8 June 2017 | 29 October 2017 | 13 March 2018 |
| End | 18 March 2017 | 29 October 2017 | 23 December 2017 | 15 August 2018 |

**Table 2.** Beach area measured at the start and end dates of the four selected periods.

| Beach Area (m$^2$) | T1 | T2 | T3 | T4 |
|---|---|---|---|---|
| Start | 99,557 | 107,674 | 99,680 | 110,841 |
| End | 108,247 | 99,680 | 113,904 | 88,914 |

In Figure 5, the shoreline positions in Hujeong Beach, $y_t$, are compared at three different times: (1) 9 December 2016, when the video monitoring data started to be measured; (2) 23 December 2017, when $A_b$ became maximum at the end time of T3; and (3) 9 September 2018, when $A_b$ was minimum at the end time of T4. It is interesting to find out that there are locations (L1, L2, L3, and L4) where $\Delta y_t$ was greater than the other areas. The reason for this non-homogeneous $y_t$ evolution along the shoreline is probably related to the nearshore sandbars and rocky seabeds that are observed at 3–5 m water depth. For example, Figure 5b shows a satellite image around the four locations from L1 to L4 on 20 April 2018 (T4), captured from Google Earth. Google Earth images are constructed based on the data from Landsat satellites. Landsat images were employed to show the shallow water seabed topography when the images are reflected through the water column under clear weather conditions [14,15]. The Google image in Figure 5 shows that crescentic sandbars actively develop in the nearshore area of the beach, specifically near L1 and L2. Besides the sandbars, underwater rocks are abundantly observed around L3 (black dots inside rectangle R1). In addition, small rocky islands and underwater rocks around the islands are found to be located near L4 within rectangle R2. These sandbars and rocky seabeds, along with the incoming waves, affected the shoreline change pattern during the periods, which will be examined in the next section.

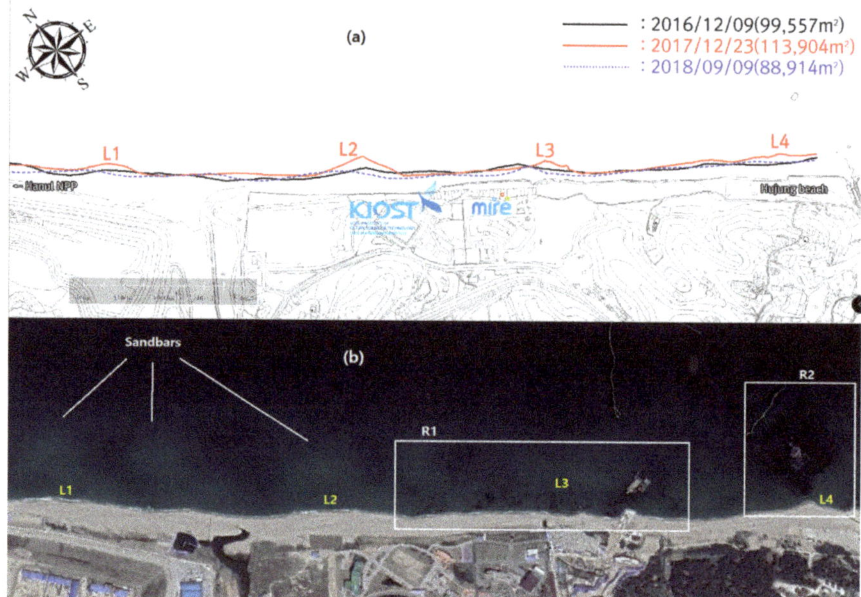


**Figure 5.** (**a**) Comparison of shoreline positions, $y_t$, measured on 9 December 2016 (start date of data collection), 23 December 2017 (when $A_b$ became maximum at T3), and 9 September 2018 (when $A_b$ was minimum at T4). L1, L2, L3, and L4 denote the locations where the shoreline positions, $y_t$, significantly changed during the period. (**b**) Satellite image on 20 April 2018 (T4), captured from Google Earth. The white rectangles, R1 and R2, denote the locations where underwater rocks develop near L3 and L4, respectively. The figure also shows the development of nearshore sandbars, indicated with the white lines.

## 4. Results

In Figure 6, $y_t$ at two times—beginning and end times of T1—are compared. The red and blue colors filled in-between the two lines denote the advance and retreat, respectively, of the shoreline during T1. In the figure, a rose diagram that shows the distribution of the propagating wave heights, $H_w$, and directions, $D_w$, is also drawn to support the understanding of the shoreline evolution pattern. $H_w$ and $D_w$ were estimated from the wave spectra based on the wave data measured by KIOST using Nortek Acoustic Waves and Currents (AWAC) at the location of ~19 m depth and ~1.0 km away from the coast. For the three-month period of T1 when $A_b$ significantly fluctuated by repeatedly losing and gaining the total beach area, $\Delta y_t$ shows an inconsistent pattern along the coastline because the shore significantly advanced at L3 and L4 while it decreased at L1 and L2. In this period, ~50% of the waves approached the shore in NNE direction (slightly inclined to the north) and ~30% of the waves approached the shore in the normal direction ($D_w$ = NE). Regarding the propagating wave heights, ~20% of them were high waves with $H_w > 2$ m. Therefore, the shoreline accretions at L3 and L4 were caused when shadow zones were formed by the obliquely approaching waves in the lee area of the underwater rocks developed in R1 and the rocky island in R2. These inclined waves could produce longshore sediment transport to the southeast direction along the beach. In the shadow zones behind the rocks, however, the sediment transport rates were reduced to cumulate sediments, specifically in L3 and L4, which can be easily understood from a simple consideration of the continuity equation for the one-line theory [16] as

$$h_p \frac{dy}{dt} + \frac{dQ}{dx} = 0 \tag{1}$$

where $h_p$ is the active profile height, $y$ is the shoreline position, and $Q$ is the longshore sediment transport. According to Equation (1), therefore, the shoreline would advance ($\Delta y > 0$) if the sediment transport rate decreases ($\Delta Q < 0$), as observed in L3 and L4. At locations near L1 and L2, $y_t$ slightly advanced to the sea. This shoreline change may be related to sandbar development as shown in Figure 5b, which will be discussed later in this section.

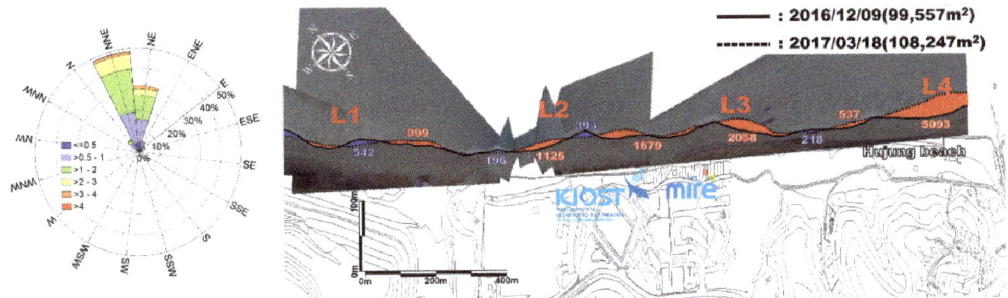


**Figure 6.** Comparison of $y_t$ between two different time steps measured at the beginning of T1 (9 December 2016) and at the end of T1 (18 March 2017) along the coastline. The red and blue colors filled in-between the two lines denote, respectively, the advance and retreat of the shoreline during T1. The numbers denote the magnitude of the gained and lost area in the corresponding colors (m²). The rose diagram shows the $H_w$ and $D_w$ distributions measured during the period.

Figure 7 compares the shoreline positions at the beginning and end time of T2 with a rose diagram of the wave conditions. It shows that the shoreline generally retreated for a ~4.5-month period by losing ~7% of the beach area, which can be estimated from Figure 4. The loss of sediments during this period is unexpected considering the relatively lower wave energy compared to T1 because ~95% of the incoming waves had $H_w$ lower than 2 m. As for $D_w$, the directional spectrum is wider compared to T1 but the majority of the waves approached the shore in the normal direction, which reduced the amount of sediments that moved in the longshore during the period. Instead, more sediments were probably transported in the cross-shore direction. For example, the shoreline at the end time of T2 became flatter compared to that at the beginning time, when the shoreline was more severely undulated. During T2, therefore, wave forces acted uniformly along the coast regardless of the existence of underwater rocks under normal direction approaching wave conditions. The rose diagram in Figure 7 shows that extreme wave conditions with $H_w > 4$ m were observed (~5% of total waves). These high waves probably contributed to the erosive process. It should be noted, however, that the beach area gradually decreased during T2 instead of having a specific time of rapid reduction in $A_b$, and thus, the impact of extreme waves on the shoreline change is still not clearly identified based on the wave data.

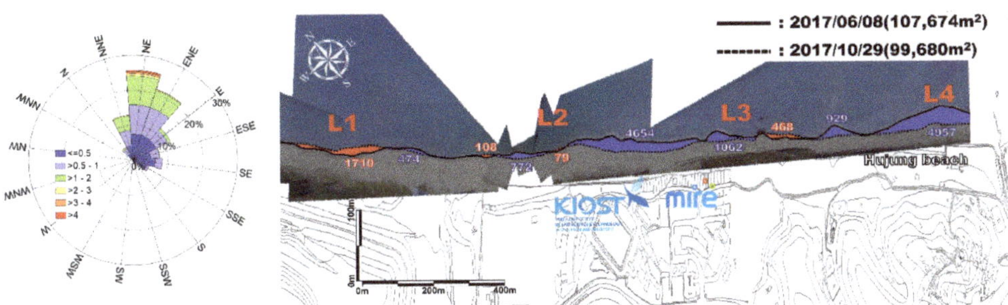


**Figure 7.** Same comparison as that in Figure 6 but measured at the beginning (8 June 2017) and end (29 October 2017) of T2.

Figure 8 shows the shoreline positions at the beginning and end time of T3 (29 October to 23 December 2017) with the corresponding rose diagram of wave conditions. It is interesting to find out that $A_b$ increased ~15% during the three-month period. Specifically, the shoreline significantly advanced at the four locations from L1 to L4. The wave conditions in T3 were similar to those in T1 as most of the incoming waves propagated obliquely from the north of the beach ($D_w$ ~NNE). Therefore, shadow zones were formed to cause the shoreline accretion at L3 and L4 where underwater rocks developed, as was observed in the case of T1. One of the differences in the shoreline evolution pattern between T1 and T3 is that the shoreline was significantly advanced at L1 and L2 in T3, which was not observed in T1. For example, $\Delta y_t$ became maximum at L2 and it was also dominant at L1 in T3, while $\Delta y_t$ showed no clear changes at these locations in T1. As already mentioned, the significant shoreline changes at L1 and L2 were related to the development of nearshore sandbars.

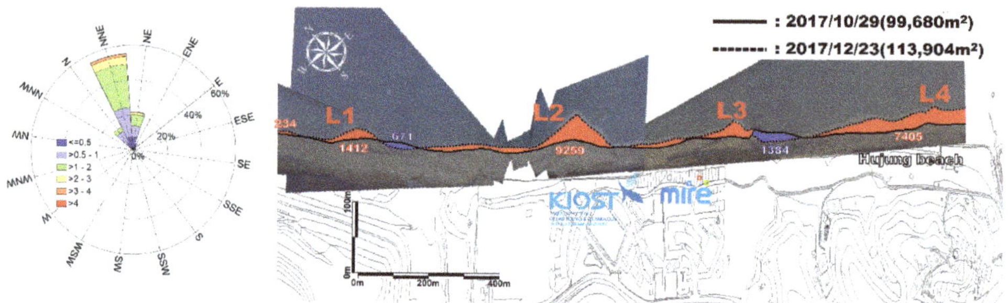


**Figure 8.** Same comparison as that in Figure 6 but measured at the beginning (29 October 2017) and end (23 December 2017) of T3.

Figure 9 shows the seabed elevation changes in Hujeong Beach between 18 November 2016 (before T1) and 12 March 2017 (after T1) [17]. The bottom topography was measured using echosounders and a mobile LiDAR. The changes in seabed elevation show that nearshore sandbars actively developed parallel to the shoreline, in the area with water depth shallower than 5 m. The development of sandbars can be confirmed with the Google image shown in Figure 5b because crescentic sandbars are observed near the locations of L1 and L2 on 20 April 2018 (T4). The figure also shows that the shoreline is advanced to the offshore at the locations where the horns of the sandbars developed near L1 and L2, although the underwater rocks do not develop in these areas. It should be noted that these locations of the shoreline accretion correspond to the locations where the horns of the sandbars developed, which shows a good example of out-of-phase coupling between the sandbars and shoreline [18,19]. Therefore, the dominant shoreline accretions at L1 and L2 indicate that nearshore crescentic sandbars might be actively developed at T3, and the shoreline was advanced at the locations corresponding to the sandbar horns. Compared to T1, the wave energy in T3 was lower as the waves with $H_w > 2$ m were ~7% of total incoming waves in this period, which might provide favorable conditions to form crescentic sandbars as a result of positive feedback between the nearshore flow circulation, sediment transport, and the morphology itself [18].

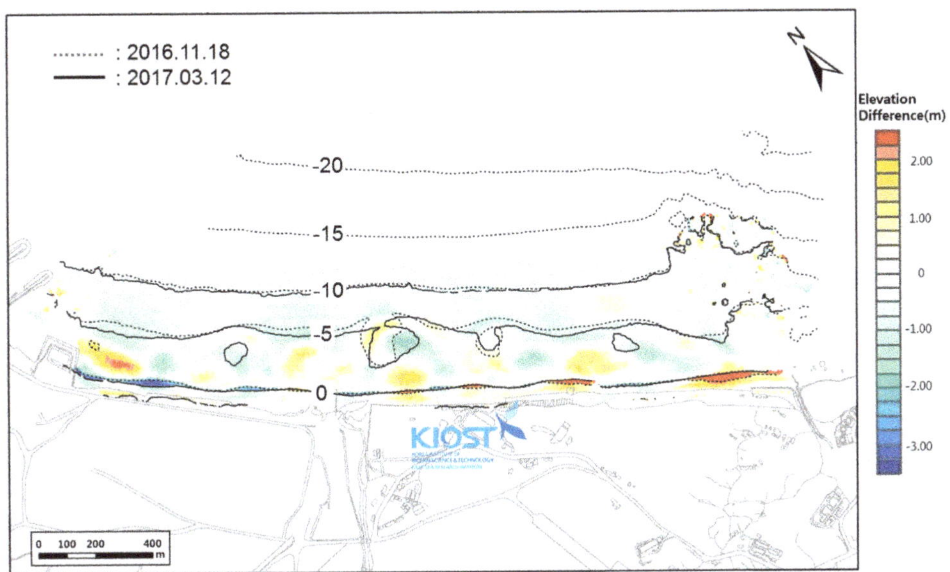


**Figure 9.** Seabed elevation changes in Hujeong Beach between 18 November 2016 (before T1) and 12 March 2017 (after T1). The figure is taken from [17].

Figure 10 shows another example of significant shoreline change between the beginning and end time of T4 (13 March to 15 August 2018), when $A_b$ decreased ~19%. The shoreline was generally retreated along the entire coast, but the reduction in $y_t$ was particularly greater at L2 and L4. As shown in Figure 9, the shoreline accretion at L1 and L2 during T3 was due to the nearshore sandbars. Therefore, the shoreline retreat at the corresponding locations might also be related to the sandbars, which could occur when the sandbar development became weaker during T4. It should be noted that the wave conditions in T4 can be distinguished from those in periods T1–T3. The wave conditions in T4 were generally similar to those in T3 because the wave energy was moderate as ~10% of $H_w$ was higher than 2 m. However, compared to the other cases, $D_w$ spread in wider angles, showing that the waves were approaching in various directions. Interestingly, ~20% of $D_w$ were SE and ESE, and this indicates that the waves approached parallel to the shore from the south. These shore-parallel waves might significantly affect the shoreline changes by transporting the sediments along the shoreline regardless of the relatively low wave energy ($H_w$ < 1 m).

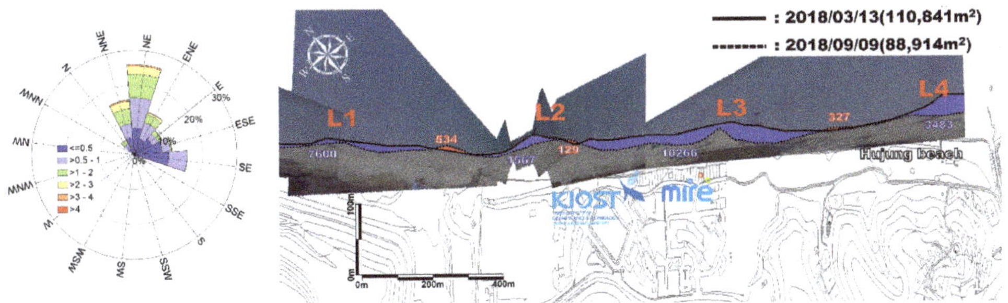


**Figure 10.** Same comparison as that in Figure 6 but measured at the beginning (13 March 2018) and end (15 August 2018) of T4.

One evidence on the longshore transport of the sediments near the shore during T4 is found in Figure 11, in which the Google images near L1 and L2 are compared at two times, on 3 March 2018 (T4) and 20 April 2018 (T4). The two times of the captured images belong to the early stage of T4, and they could provide an implication of the littoral process in this period. In Figure 11a,b, the yellow lines mark the positions of the two sandbars as they were drawn to connect the horns and bays of the sandbars. Therefore, these lines compare the orientations of the sandbar development at the two times separated by ~1.5 months. On 3 March 2018 (T4), the direction of the yellow lines was generally normal to the shoreline. On April 20 (T4), however, the lines were inclined to the north and the angle deviated from the shore at normal direction was ~40°. This indicates that the sediments piled in the sandbars moved to the north due to the shore-parallel waves, which resulted in the sandbar migration along the shore. This migration of sandbars might break the equilibrium status of the out-of-phase coupling of the bars with the shoreline. Thus, the shoreline advance at the corresponding locations of the sandbar horns during T3 became weaker, which might lead to the flatter shoreline at the end time of T4. At L4, where there is a rocky island instead of sandbars, the shoreline was also significantly retreated probably due to the same shore-parallel waves that carried the sediments along the shore, regardless of the shadowing effect of the island.

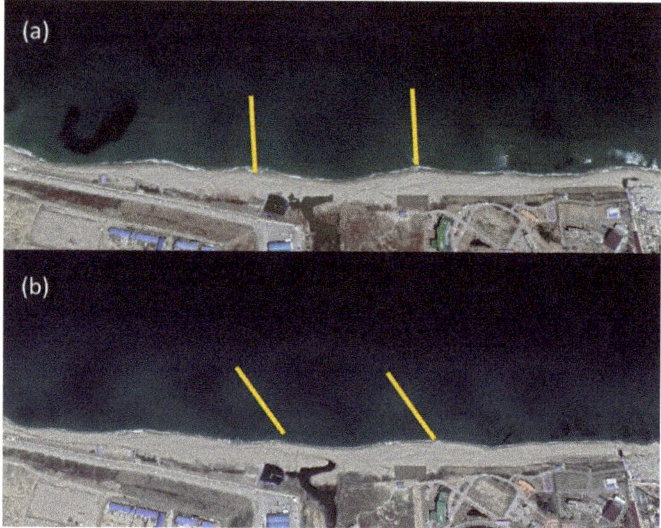


**Figure 11.** (a) Satellite image captured from Google Earth near L1 and L2 measured on 3 March 2018. (b) Same location but measured on 20 April 2018. The yellow lines denote the orientation of the horns of the sandbars.

## 5. Discussion

The shoreline data provided by the video monitoring shows that ~10% of $A_b$ was lost during the experimental period. The loss of the total beach area indicates that the beach was, in general, under an erosional process, probably due to the blocking of sediments by the breakwater near Bugu River (Figure 1). It should be noted, however, that construction of the breakwater was completed in 1989. Considering that the shoreline data was obtained 27 years after the breakwater construction, the two-year time variation of $A_b$ in Figure 4 is not sufficient to confirm that the beach is seriously erosive, and additional data may be required to support the process.

As shown in Figure 9, the seabed topography was occasionally measured using echosounders in Hujeong Beach between July 2016 and July 2018. By comparing the two bathymetry data, the changes in seabed elevation could be calculated to obtain the volume changes in the seabed. The result shows that a total of ~99,000 m$^3$ of sand were lost in the nearshore area of the beach during the period. This significant loss of sediment from the nearshore area in Hujeong Beach was continuously observed from the bathymetry data measured at other times, which supports that the beach was in an erosional process probably due to the blocking of sediments by the breakwater. In addition, Figure 12 shows an aerial photograph taken on 28 October 1980 when the intake breakwater near the Bugu River had not been constructed yet. It clearly shows that the waves obliquely approached the shore from the left side of the beach, which probably transported the sediments to the right. However, the sediments in the beach were likely in equilibrium as they were provided through the river. The equilibrium could have been broken after the breakwater was built. The shoreline positions obtained from Figure 12 implies that the beach width might have been reduced by ~15 m on average since 1980.

**Figure 12.** Aerial photograph over the Hujeong Beach area, taken on 28 October 1980 before the breakwater was constructed.

Although the beach was losing the sediments in general, the change in shoreline did not occur uniformly along the coast. Instead, there were locations where the shoreline significantly protruded to make it undulated. This process was closely related to the bottom topography formed by the nearshore sandbars and the underwater rocks that developed at nearshore areas along the beach. Figure 13 shows a satellite image from Google Earth near L1–L3 on 13 February 2019 (after T4). Although the time when the image was taken did not match with the time of T3, it still is a good example that implies the shoreline accretion process during moderate wave conditions. It can be observed in the figure that five crescentic sandbars developed along the shore around L1, L2, and L3. Interestingly, the waves were breaking at the same locations of the sandbar horns. It should also be noted that the shoreline was significantly advanced to the sea at the corresponding locations of the horns, which confirms the out-of-phase coupling of the sandbars with the shoreline. When the waves broke at horn locations due to shoaling, the wave energy in the lee area was reduced to cause a sediment transport gradient in the longshore direction. According to Equation (1), therefore, the shoreline would advance to the sea at these locations. A similar process might be expected in the area near L3 where underwater rocks actively develop. In the rocky area where the water depth was relatively shallower, the waves were easily breaking to cause a longshore sediment transport gradient and the consequent shoreline changes. This is confirmed in Figure 13, which shows that the waves were breaking, and the shoreline protruded at L3 where underwater rocks actively developed.

**Figure 13.** Satellite image captured from Google Earth near L1–L3 measured on 13 February 2019. The image shows that the waves are breaking at the locations where the sandbar horns developed, and the shoreline was advanced at the corresponding locations.

## 6. Conclusions

The shoreline evolution pattern of a sandy beach was investigated using a video monitoring system and wave measurements in Hujeong Beach, Korea. The beach is suspected to be vulnerable to erosion because the input source of sediments has been blocked by construction of a breakwater at the northern end, while the beach continues to lose sand at the opposite end. The beach is also characterized by crescentic sandbars that actively develop in the northern part of the nearshore region. In the other side of the beach, the seabed is covered by numerous underwater rocks. In addition, a rocky island is located at the southern end of the beach. These sandbars and rocky structures played significant roles in shaping the beach shorelines in concert with the incoming waves. During the two-year period of the observation, the overall area of the beach decreased ~10%. However, there were times when the beach area rapidly increased, which made the time variation pattern of beach evolution highly complicated. The erosion or accretion of the shoreline generally agreed with the expected situation based on the wave conditions. However, the response of the beach to a short-term storm event such as Typhoon Kong Rey was not clearly recorded. The video monitoring data was difficult to measure owing to the harsh weather during the storm.

The results showed that the beach area increased when the waves obliquely approached the shore, whereas it decreased when the waves approached in a normal direction to the shore, even though the wave energy was not clearly differentiated between them. In addition, the accretion of the shoreline did not occur uniformly along the beach. Instead, there were locations where the shoreline significantly protruded toward the sea. These areas corresponded to the locations where the horns of sandbars developed, which showed an example of out-of-phase coupling between the sandbars and the shoreline. As an evidence, the waves were observed to break at the same locations as the horns, which indicates that the wave energy in the lee areas of the horns was reduced to produce a longshore sediment transport gradient, resulting in shoreline accretion. Along with the sandbars, underwater rocks developed in the southern part of the beach. These rocks played a similar role to that of the sandbars, breaking the waves to produce a sheltering area in the lee. When the waves approached the shore in a normal direction, the longshore drift of sediment became weaker and offshore sediment transport occurred uniformly along the coast, which made the shoreline to become flatter. The results of this study indicate that the incoming wave directions are important in shaping the sandy shorelines, and this must be considered when designing beach facilities and coastal structures for various purposes. In addition, the seabed topography and the distribution of bedforms should be carefully investigated as they may also affect the shoreline evolution pattern. In this study, the examples of in-phase coupling between the sandbars and the shoreline have not been observed, which is expected to occur under different wave conditions and will be investigated in future studies.

**Author Contributions:** The contributions of each author are as follows: conceptualization, Y.S.C., J.-D.D., W.M.J. and J.-Y.J.; methodology, J.-D.D. and C.H.K.; formal analysis, J.-D.D. and Y.S.C.; investigation, J.-D.D. and Y.S.C.; resources, J.-D.D., W.M.J., C.H.K. and J.-Y.J.; data curation, J.-D.D.; writing—original draft preparation, J.-D.D. and

Y.S.C.; writing—review and editing, J-.D.D. and J.-Y.J.; visualization, J.-D.D. and Y.S.C.; project administration, Y.S.C. and J.-Y.J.; funding acquisition, Y.S.C. and J.-Y.J.

**Funding:** This research was funded by the Korea Institute of Ocean Science and Technology (KIOST), grant number PE99731.

**Conflicts of Interest:** The authors declare no conflict of interest.

## References


1. Lippmann, T.C.; Holman, R.A. Quantification of sand bar morphology: A video technique based on wave dissipation. *J Geophys. Res.* **1989**, *94*, 9951011. [CrossRef]
2. Plant, N.G.; Holman, R.A. Intertidal beach profile estimation using video images. *Mar. Geol.* **1997**, *140*, 1–24. [CrossRef]
3. Aarninkhof, S.G.; Turner, I.L.; Dronkers, T.D.; Caljouw, M.; Nipius, L. A video-based technique for mapping intertidal beach bathymetry. *Coast. Eng.* **2003**, *49*, 275–289. [CrossRef]
4. Davidson, M.A.; Aarninkhof, S.G.J.; Van Koningsveld, M.; Holman, R.A. Developing Coastal Video Monitoring Systems in Support of Coastal Zone Management. *J. Coast. Res.* **2006**, *1*, 49–56.
5. Chickadel, C.; Holman, R.A. Measuring longshore currents with video techniques. In *EOS Trans, Fall Meeting*; American Geophysical Union: Washington, DC, USA, 2002; Volume 82.
6. Holmann, R.A.; Stanley, J. The history and technical capabilities of Argus. *Coast. Eng.* **2007**, *54*, 477–491. [CrossRef]
7. Yoon, J.; Song, D. Preliminary study for detecting of ocean wave parameters using CCD images. *J. Coast. Res.* **2018**, *85*, 1371–1375. [CrossRef]
8. Ojeda, E.; Guillén, J. Monitoring beach nourishment based on detailed observations with video measurements. *J. Coast. Res.* **2006**, *48*, 100–106.
9. Vousdoukas, M.I.; Almeida, L.P.M.; Ferreira, Ó. Beach erosion and recovery during consecutive storms at a steep-sloping, meso-tidal beach. *Earth Surf. Process. Landf.* **2012**, *37*, 583–593. [CrossRef]
10. Holland, K.T.; Holman, R.A.; Lippmann, T.C.; Stanley, J.; Plant, N. Practical use of video imagery in nearshore oceanographic field studies. *IEEE J. Ocean. Eng.* **1997**, *22*, 8192. [CrossRef]
11. Chang, Y.S.; Do, J.-D.; Kim, S.-S.; Ahn, K.; Jin, J.-Y. Measurement of turbulence properties at the time of flow reversal under high wave conditions in Hujeong Beach. *J. Korean Soc. Coast. Ocean Eng.* **2017**, *29*, 206–216. [CrossRef]
12. Korea Institute of Ocean Science and Technology (KIOST). Development of fundamental technology for coastal erosion control. *Tech. Rep.* **2018**, *5*, 43–52.
13. Zhang, Y.; Wetherill, B.R.; Chen, R.F.; Peri, F.; Rosen, P.; Little, T.D.C. Design and implementation of a wireless video camera network for coastal erosion monitoring. *Ecol. Inform.* **2014**, *23*, 98–106. [CrossRef]
14. Jagalingam, P.; Akshaya, B.J.; Hedge, A.V. Bathymetry mapping using Landsat 8 satellite imagery. *Procedia Eng.* **2015**, *116*, 560–566. [CrossRef]
15. Pacheo, A.; Horta, J.; Loureiro, C.; Ferreira, Ó. Retrieval of nearshore bathymetry from Landsat 8 images: A tool for coastal monitoring in shallow waters. *Remote. Sens. Env.* **2015**, *159*, 102–116. [CrossRef]
16. Pelnard-Considere, R. Essai de Theorie de l'Evolutio des Form de Rivage en Plage de Sable et de Galets. *4th Journees de l'Hydraulique, Les Energies de la Mer, Question III* **1956**, *1*, 289–298.
17. Korea Institute of Ocean Science and Technology (KIOST). Advanced Operation of the Coastal Hydrodynamics and Beach Erosion Monitoring System Infra in Hujeong. *Tech. Rep.* **2019**, *7*, 57–65.
18. Price, T.D. Morphological Coupling in a Double Sandbar System. Ph.D. Thesis, Utrecht University, Utrecht, The Netherland, 2013.
19. Price, T.D.; Ruessink, B.G.; Castelle, B. Morphological coupling in multiple sandbar systems—A review. *Earth Surf. Dyn.* **2014**, *2*, 309–321. [CrossRef]

 *applied sciences*

*Article*

# Wind-Driven Hydrodynamics in the Shallow, Micro-Tidal Estuary at the Fangar Bay (Ebro Delta, NW Mediterranean Sea)

**Marta F-Pedrera Balsells *, Manel Grifoll, Manuel Espino, Pablo Cerralbo and Agustín Sánchez-Arcilla**

Maritime Engineering Laboratory (LIM), Catalonia University of Technology (UPC), 08034 Barcelona, Spain; manel.grifoll@upc.edu (M.G.); manuel.espino@upc.edu (M.E.); pablocerralbo@gmail.com (P.C.); agustin.arcilla@upc.edu (A.S.-A.)
* Correspondence: marta.balsells@upc.edu

Received: 18 August 2020; Accepted: 1 October 2020; Published: 4 October 2020

**Abstract:** This article investigates water circulation in small-scale (~10 km$^2$), shallow (less than 4 m) and micro-tidal estuaries. The research characterizes the hydrodynamic wind response in these domains using field data from Fangar Bay (Ebro Delta) jointly with three-dimensional numerical experiments in an idealized domain. During calm periods, field data in Fangar Bay show complex water circulation in the inner part of the estuary owing to its shallow depths and positive estuarine circulation in the mouth. Numerical experiments are conducted to investigate wind-induced water circulation due to laterally varying bathymetry. For intense up-bay wind conditions (wind intensities greater than 9 m·s$^{-1}$), an axially symmetric transverse structure occurs with outflow in the central channel axis and inflow in the lateral shallow areas. These numerical results explain the water circulation observed in Fangar Bay during strong wind episodes, highlighting the role of the bathymetry in a small-scale environment. During these episodes, the water column tends to homogenize rapidly in Fangar Bay, breaking the stratification and disrupting estuarine circulation, consistent with other observations in similar domains.

**Keywords:** Fangar bay; wind; shallow and small-scale bay; estuary; micro tidal

---

## 1. Introduction

Water circulation induced by local wind effects can contribute to the water exchange between an estuary and the adjacent open sea [1]. Seminal investigations presented by Csanady [2] described how wind stress applied at the surface of a basin of variable depth sets up a circulation pattern characterized by coastal currents in the direction of the wind, with return flow occurring over the deeper regions. Narváez et al. [3] showed that the observed transverse partition of the subtidal circulation is consistent with that driven by local wind in a channel with lateral depth variations. Wong et al. [4] showed how the impact of the wind-induced coastal sea level at the entrance of an inlet exerts a huge impact on the subtidal variability within the interior of the bay. The influence of wind can also have effects on salinity structure or resuspension events in shallow estuaries [5–8]. These studies indicate that spatial scales and water depth determine the flow response to wind forcing. Moreover, Noble et al. [9] determined that both river and wind-driven flow patterns change as a function of water depth in a shallow estuary (below 4 m). In this case, for wind-driven flows the critical parameter is the degree of stratification in the lower bay. However, the particular role that plays the geometry and the bathymetry in wind response in a small-scale bay is still not clear. An illustrative contribution presented by Sanay and Valle-Levinson [10] reflects the role of the bathymetry in the water contribution for larger and deeper basins than that mentioned in this study. Numerical experiments reveal a wind-induced water circulation dominated by an axially symmetric transverse structure due to laterally varying bathymetry. So, the investigation at very small-scales (order of few km) and shallow depths (few meters) still presents challenges that may be addressed by complementing observational and numerical efforts. The short time scales expected due to the shallowness, the role that plays the geometry (for instance the mouth width) and the link between bathymetry and water circulation are challenging questions to describe properly small-scale bays under micro-tidal conditions.

Contributions focusing on coastal bays and estuaries conclude that they are regions with complex hydrodynamics due to multiple forcing [7,9,11]. In some cases, wind stress competes with baroclinicity to determine the hydrodynamics. However, the astronomic tide [12], seiches [13], water run-off, the bathymetry or the effect of rotation [10] are variables that may not be neglected in the analysis of the local wind influence on this type of water bodies. In these cases, the combination on observational and numerical efforts have allowed obtaining substantial new knowledge on water circulation and termohaline structure.

This paper investigates the relationship between wind and hydrodynamics of small-scale, shallow and micro-tidal estuary. The research combines "in situ" data using the case study of Fangar Bay, located in the Ebro delta, NW Mediterranean Sea, and the implementation of numerical experiments in idealized geometries. These simulations should figure out the impact of the bathymetry on wind-driven water circulation. The complexity of the flow observed have suggested to focus, at this first stage, on the effect of the bathymetry in the wind-driven circulation. Fangar Bay presents problems of water renewal, high water temperatures, anoxia and permanence of pollutants in the water. All these problems affect the oyster and mussel crops present in the bay that need a sustained water renewal for their growth and development [14]. Several intensive field campaigns have been carried out to characterize the bay hydrodynamics as a first step for a subsequent water renewal analysis. In this sense, Fangar Bay is a good example to investigate the hydrodynamics of small-scale and shallow scale bays under micro-tidal conditions.

## 2. Materials and Methods

### 2.1. Study Area

Fangar Bay is part of the Ebro Delta (NW Mediterranean Sea), which extends around 25 kilometres offshore and forms two enclosed bays (Fangar to the north and Alfacs to the south). Both receive freshwater discharges from the irrigation channels of rice fields. The region is micro-tidal and previous analysis in these bays have observed a negligible influence of the tide on the water circulation [5,12]. The main dimensions of Fangar Bay are 12 km$^2$, with a length of about 6 km, a maximum width of 2 km and a volume of water of $16 \times 10^6$ m$^3$ [15]. The average depth is 2 m, with a maximum of 4 m (see bathymetry in Figure 1). The connection with the open sea is located NW and spans approximately 1 km [16]. In recent years, the mouth width has been modified by the accumulation of sediment from the beach located to the north [17], so its width is currently less than 1 km.

The Ebro Delta is influenced by the presence of southerly/south-easterly sea breezes—which do not exceed 6 m·s$^{-1}$ during the summer and spring seasons—and strong winds from the north and west of more than 12 m·s$^{-1}$ in autumn and winter [18,19]. The most frequent wind throughout the year is the NW wind, locally known as the mestral, which is characterized by strong gusts of cold and dry wind [20]. In addition, sporadic easterly winds occur, associated with rainy events [21].

The freshwater contribution is regulated by rice cultivation throughout the year. Open channels occur from April to December receiving a mean flow of 7.23 m$^3$·s$^{-1}$ [22]. This flow is distributed irregularly throughout the year, with maximum values in the months of April to November. Negligible freshwater flow occurs from December to March when the channels are closed [23]. The freshwater flows during these fields campaigns is estimated in 6 m$^3$·s$^{-1}$. This flow is distributed in two main freshwater discharges in Fangar Bay: one in the Illa de Mar harbour and the other, Bassa de les Olles, located in the bay mouth (Figure 1). Additional freshwater discharges along the coastline are expected because freshwater inputs are regulated by gravity according to sea level. Finally, a substantial contribution from groundwater inputs is expected within the bay [15]. In both cases, the expected freshwater flow is smaller than the water pumping stations mentioned above.

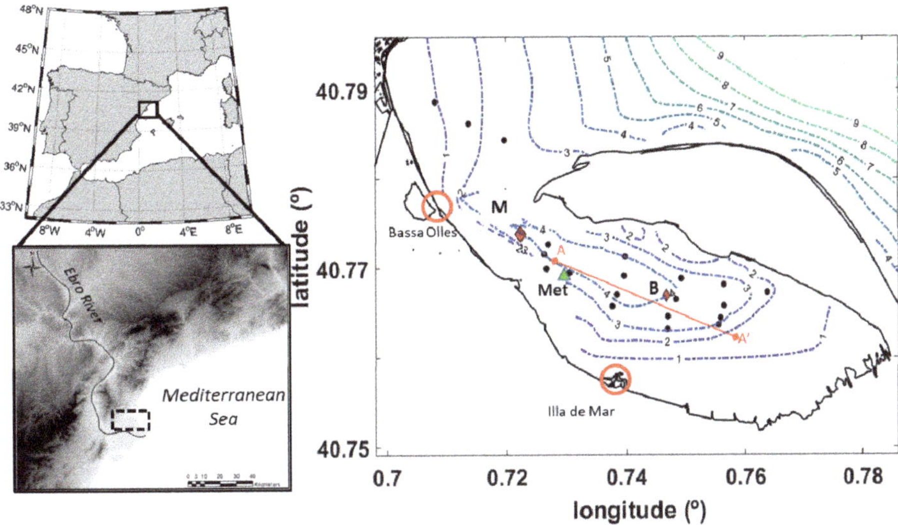

**Figure 1.** Location of the study area. The red circles show the two main points of freshwater discharges (Bassa de les Olles and Illa de Mar), the red diamonds show the points of measurement during the field campaigns (M: mouth; B: bay) and the green triangle shows the location of the meteorological station (Met) during FANGAR-II. The black point shows the conductivity, temperature and depth (CTD) profiles, the blue lines show the bathymetry and the red line (A–A′) shows the longitudinal section presented in Figure 9.

### 2.2. Field Campaigns in Fangar Bay

The observational data corresponded to two field campaigns (each of around two months) from 25 July to 5 September 2017 (FANGAR-I) and from 5 October to 16 November 2017 (FANGAR-II), hence summer and autumn, respectively. The data set consisted of two Acoustic Doppler Current Profilers (ADCPs, from NORTEK, model AQUADOPP 2 MHz) (mooring points M and B in Figure 1), with the velocity and the direction of the water currents obtained every 10 min in 25 cm layers distributed from the bottom to the surface. Moreover, the systems were equipped with pressure systems and a temperature sensor (Vaisala PTB110 and HMP40). A set of CTD (conductivity, temperature and depth, model SeaBird 19plus) campaigns were conducted. During FANGAR-I, 16 CTD campaigns were carried from 11 July to 5 September (two campaigns per week). During FANGAR-II, only two surveys were undertaken, one at the beginning of the campaign (18 October) and the other at the end (16 November). For each of the CTD campaigns, twenty points were chosen, including both the inner and the outside sections of the bay, where temperature and salinity were measured (see black dots in Figure 1).

During FANGAR-I, data from Servei Meteorologic de Catalunya (MeteoCat, http://www.meteo.cat/ (accessed on 30 January 2020)) of the Illa de Buda, located about 12 km from Fangar Bay, were used. This station records wind data every 30 min. During FANGAR-II, a meteorological station (with temperature and humidity sensor, air pressure and acoustic wind sensor) was mounted on a mussel raft near the mouth (Met in Figure 1) to measure wind, atmospheric pressure, air temperature and humidity every 10 min. The measurement periods and instruments are summarized in Table 1.

**Table 1.** Data acquisition instruments and observational periods (year 2017) shown in Figure 1.

| Name (ID) | Observations | Period | Data Interval (min) |
| --- | --- | --- | --- |
| Meteo station (Met) (wind Sonic/Vaisala HMP40/Vaisala PTB110) | Wind, atmospheric pressure | 19 October–16 November | 10 |
| Illa de Buda station (wind Sonic/Vaisala HMP40/Vaisala PTB110) | Wind, atmospheric pressure | 25 July–5 September | 30 |
| ADCP mouth (M) (Nortek Aquadopp 2 MHz) | Currents, sea level, waves, bottom temperature | 25 July–5 September<br>5 October–16 November | 10 |
| ADCP inner bay (B) (Nortek Aquadopp 2 MHz) | Currents, sea level, waves, bottom temperature | 25 July–5 September<br>5 October–16 November | 10 |
| CTD (SeaBird 19plus) | Temperature, salinity | 11 July–5 September | - |
|  |  | 18 October and 16 November | - |

## 2.3. Numerical Modeling

A series of numerical experiments were conducted using the Regional Ocean Model System (ROMS) to analyse the hydrodynamic wind response in small and shallow estuaries. The ROMS numerical model is a 3D, free-surface, terrain-following numerical model that solves the Reynolds-averaged Navier–Stokes equations using hydrostatic and Boussinesq assumptions [24]. ROMS uses the Arakawa-C differentiation scheme to discretize the horizontal grid in curvilinear orthogonal coordinates and finite difference approximations on vertical stretched coordinates [25]. The numerical details of ROMS are described extensively in [24]. This model has been used and validated in similar bays and estuaries, such as Alfacs Bay located south of the Ebro Delta (e.g., [13,26,27]). The domain used is inspired in Fangar Bay and consists in an idealized domain due to the difficulty of achieving good bathymetry. The domain consists of a regular 37 × 27 grid with a horizontal resolution of about 70 m (Figure 2a) and 10 sigma levels in the vertical direction (the bottom layer being the first and the surface layer the tenth, Figure 2b). The model boundary is located 10 points away of the mouth entrance to avoid boundary noise (faster than show in the Figure 2a). The 2D variables were accommodated with the Chapman and Flather algorithms [28], whereas a clamped boundary condition is imposed on the 3D variables. The same model configuration has been used and validated in the bay located south of the Delta, the Alfacs Bay [27].

Figure 2a presents the different geometries used for the numerical experiments. We modeled a simple case, regarding the water body as a channel 2.5 km long and 1.9 km wide, to simplify the response of the wind to the idealized bathymetry similar to that of the Fangar Bay. The first was of a homogeneous depth of 4 m (GEO 1). The second was variable bathymetry (v-shape channel with 1 m to the sides up to 4 m in the central part, GEO 2). The same experiment was then conducted, but reducing the mouth of the channel to simulate the narrow entrance of the Fangar Bay (GEO 3). Finally, it was simulated to include variable bathymetry (GEO 4). Experiments were conducted up-bay wind and down-bay wind according to the axis of the estuary (wind spatially homogenous). An additional simulation took into account the Coriolis effect for GEO 1 to verify the influence of the Earth's rotation. All simulations are summarized in Table 2. Constant temperature (20 °C) and salinity (36) conditions were applied. Freshwater inputs were deactivated in order to observe the behaviour of the currents only as a function of wind and bathymetry. The bottom boundary layer was parameterized with a logarithmic profile using a characteristic bottom roughness height of 0.002 m. The turbulence closure scheme for the vertical mixing was the generic length scale (GLS) tuned to behave as a k-$\varepsilon$ [29]. Horizontal harmonic mixing of momentum was defined with constant values of 5 $m^2 \cdot s^{-1}$. Long-term simulations (around 2 months each) were performed in order to analyze the wind response avoiding spurious velocities. The wind is constant throughout the simulation. According to the results less 24 h of the spin-up is observed in the numerical results. The numerical results used are 16 days after the start of the simulation to avoid spurious velocities.

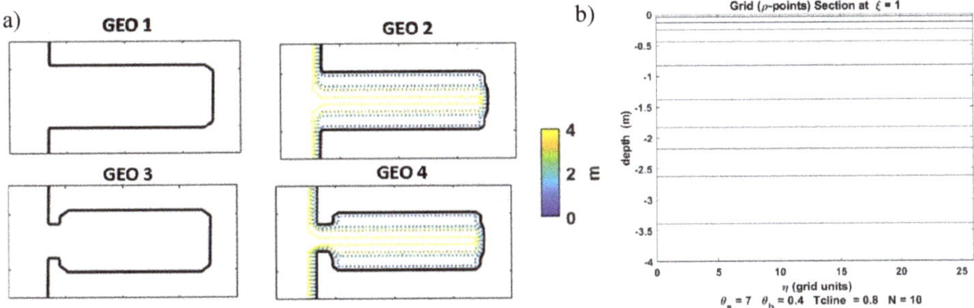

**Figure 2.** (a) Geometry and bathymetry for the four configurations used in the numerical experiments. GEO 1 and GEO 3 have 4 m of water depth. (b) Sigma layers used in the numerical model. The geometrical parameters are included following the stretching formulation described in Cerralbo et al. (2015).

**Table 2.** Summary of the different simulations used in Regional Ocean Model System (ROMS).

| Simulation Name | Geometry | Wind Direction | Intensity Wind (m·s$^{-1}$) | Coriolis |
|---|---|---|---|---|
| SIMU 1 | GEO 1 | Up-bay wind | 12 | No |
| SIMU 2 | GEO 2 | Up-bay wind | 12 | No |
| SIMU 3 | GEO 2 | Up-bay wind | 6 | No |
| SIMU 4 | GEO 2 | Down-bay wind | 12 | No |
| SIMU 5 | GEO 3 | Up-bay wind | 12 | No |
| SIMU 6 | GEO 4 | Up-bay wind | 12 | No |
| SIMU 7 | GEO 4 | Down-bay wind | 12 | No |
| SIMU 8 | GEO 1 | Up-bay wind | 12 | Yes |

## 3. Results

### 3.1. Meteorological Description of Fangar Bay

Wind roses for both field campaigns are shown in Figure 3. During FANGAR-I the predominant winds were southerly and easterly, associated with sea breezes, whereas during FANGAR-II the predominant and strongest winds were from the NW (i.e., an upward wind direction considering the axis of the estuary). Note the presence of NW winds in both campaigns.

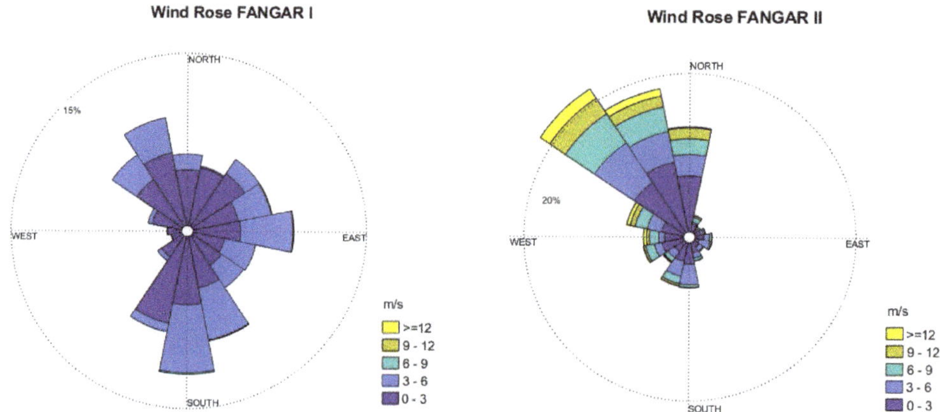



**Figure 3.** Wind roses of mean velocity for FANGAR-I (**left**) and FANGAR-II (**right**) during both campaigns. Wind intensities are grouped by intervals of 3 m·s$^{-1}$.

Figure 4 shows the time series of the wind and air temperature measured at the corresponding meteorological station. During summer (FANGAR-I), the sea breeze was characterized by daily southerlies with a wind intensity of 6 m·s$^{-1}$. Furthermore, two episodes of NW winds were distinguished during FANGAR-I—from 6 to 12 August (henceforth called episode E1) and from 29 August to 3 September (henceforth called E2)—as well as in FANGAR-II from 21 to 23 October (henceforth called E3) and from 5 November until the end of the campaign (henceforth called E4). These NW episodes could also be identified alongside the air temperature drops in Figure 4 superimposed at daily variability during both field campaigns.

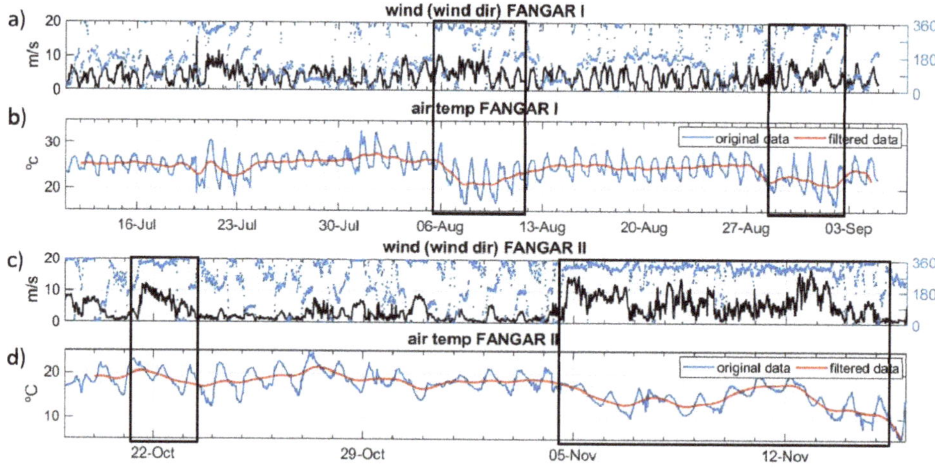



**Figure 4.** Wind direction and intensity from FANGAR-I (**a**) and FANGAR-II (**c**). Air temperature for FANGAR-I (**b**) and FANGAR-II (**d**). Air temperature is filtered to exclude daily variability. The black box shows the periods with falling air temperatures during NW winds.

### 3.2. Hydrodynamic Description of Fangar Bay

A dispersion diagram of the water currents (surface and bottom) measured by the ADCPs in each campaign is shown in Figure 5. The diagram shows how in the mouth (M station) the flow was aligned following the longitudinal axis of the bay. By contrast, in the B position the flow did not show a prevalent pattern, the water circulation being more complex. In addition, Figure 5 demonstrates how the water flow in the mouth was larger than in the inner bay. The maximum water currents measured in the mouth were 0.5 m·s$^{-1}$ in FANGAR-I and FANGAR-II, but below 0.02 m·s$^{-1}$ in both campaigns within the bay. Furthermore, the surface and the bottom signal showed similar patterns across the campaigns.

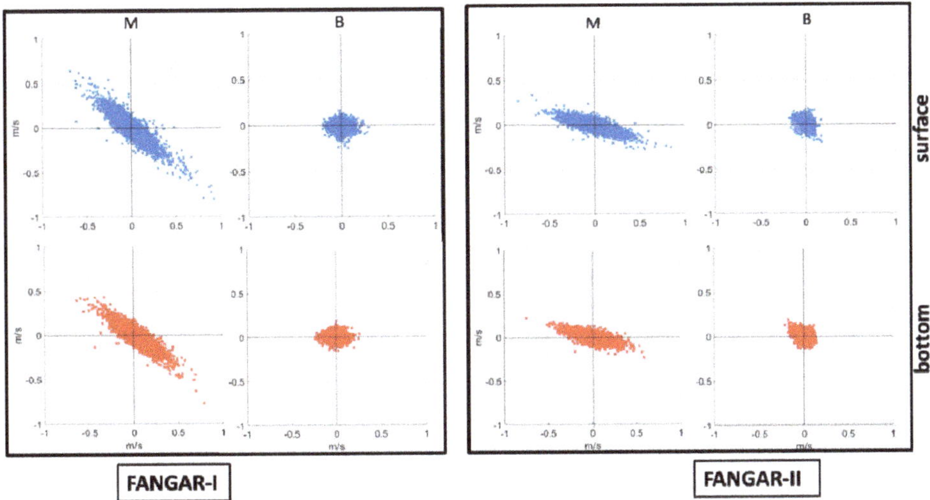


**Figure 5.** Dispersion diagram for FANGAR-I (**left**) and FANGAR-II (**right**), in blue the surface water circulation and in red the bottom water circulation. The first column of panel corresponds to the measurements in the mouth (M) and the second column to those within the bay (B).

Regarding the flow alignment shown in the dispersion diagram (see Figure 5), the water current observations were rotated following the alongshore (longitudinal axis of the bay with positive inward) and cross-shore alignment. The rotation angle was equal to 35° for FANGAR-I and 15° for FANGAR-II (north clockwise negative) according to the flow alignment of the dispersion diagram. Discrepancies in the rotation angles probably owed to differences in the mooring location of the ADCPs. Figure 6 shows the alongshore time-averaged currents for each measurement layer of the ADCPs. The resultant profiles in the water column revealed a predominant two-layer structure during FANGAR-I (Figure 6a), indicating that estuarine circulation was the surface current leaving the bay (negative values), while bottom currents were entering (positive values). During FANGAR-II, the water circulation pattern was similar to FANGAR-I, aside from differences in surface values (above 1 m depth, Figure 6b) when averaging the whole period. However, after differentiating calm wind or sea breeze periods with those of the NW winds (i.e., E3 and E4), distinctions became evident. Indeed, during such periods the profile corresponded to the positive estuarine circulation pattern, normal behaviour in this type of geometry (Figure 6c), whereas when considering only the NW wind episodes that lasted the entirety of November, the water currents were negative (leaving the bay, Figure 6d) in almost all water column broking down the positive estuarine circulation.

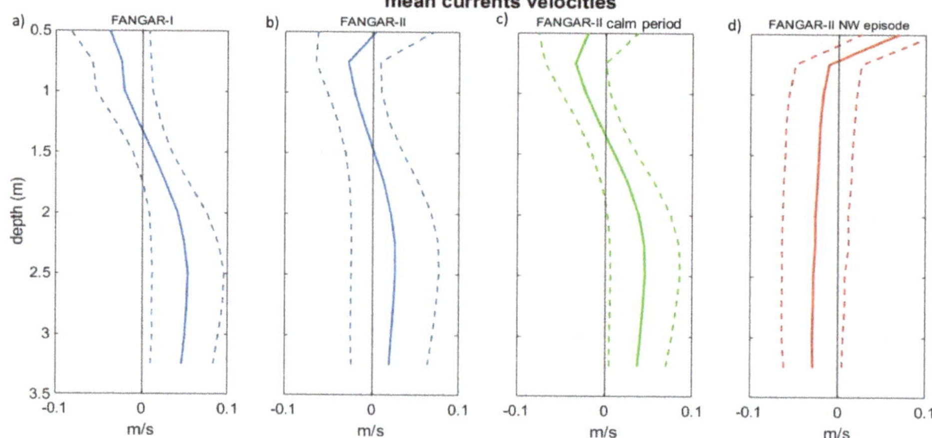



**Figure 6.** Average alongshore current velocity (positive outward) profile (continuous line) and standard deviation (dashed lines) for the mouth in FANGAR-I (**a**) and in FANGAR-II (**b**); (**c**,**d**) show the mean profile during FANGAR-II for the calm period (**c**) and the NW wind period (**d**).

Figure 7 shows the spectral analysis of the alongshore current in the mouth and in the inner bay. High-frequency peaks corresponding to the tidal ranges and typical resonance periods are shown. The seiching band using the Merian formula ($2L/n\sqrt{gH}$) approximately corresponded to periods of two hours in the fundamental mode (considering a characteristic length of about 6 km). These values are consistent with other locations in semi-closed bays [12,29,30].

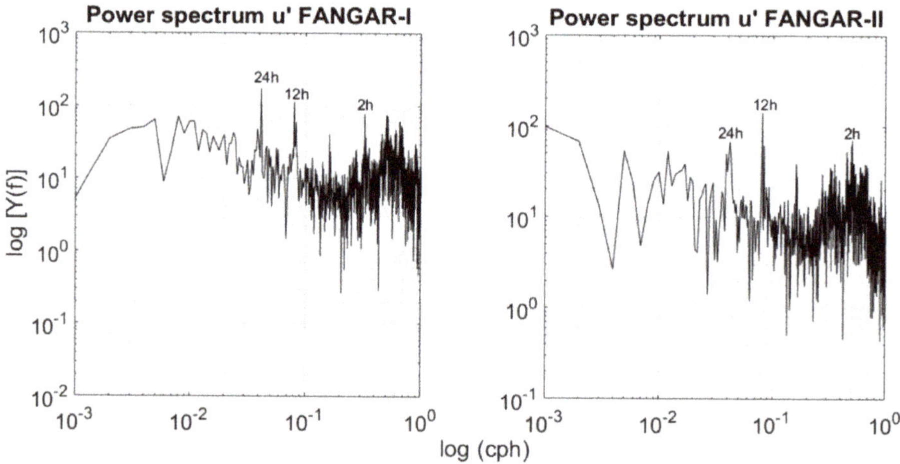



**Figure 7.** Spectral analysis of the signal of the alongshore current in the mouth for FANGAR-I (**left**) and FANGAR-II (**right**).

In order to conduct an episodic analysis, the time series were filtered using a 30 h Lanczos [31] to focus on the low-frequency signal. Figures 8 and 9 show the rotated and filtered components of water currents jointly with wind, temperature and mean flow. Additional measurements of T and S are also shown in Figure 8, with Figures 8d,e and 9c–e presenting the estuarine circulation mentioned previously, whereby the surface water flows out from the bay (negative and blue color) while the bottom water circulation enters (positive and red color). These periods were characterized by a strong stratification. This was due to the fact that the irrigation channels during FANGAR-I were open, contributing to the discharge of fresh water into the bay (Figure 8c). The same occurred at the beginning of FANGAR-II. Marked with a black box are the episodes of NW winds in Figure 8. Note that during these episodes, both temperature and salinity were homogenized (Figure 8b,c) and there was a change from positive estuarine circulation to homogeneous currents in the water column. Moreover, a substantial drop in sea bottom temperature occurred (Figure 8f), notably during episodes E1 and E2 in Figure 8, and during events E3 and E4 in Figure 9. Note that during NW wind episodes, the currents were opposite to the wind direction in both locations (i.e., M and B). Figures 8g and 9g shows the average flow through the mouth. It can be seen that during the NW episodes the net flow is outflow in the central area of the channel, while in the rest of the period occurs the opposite.

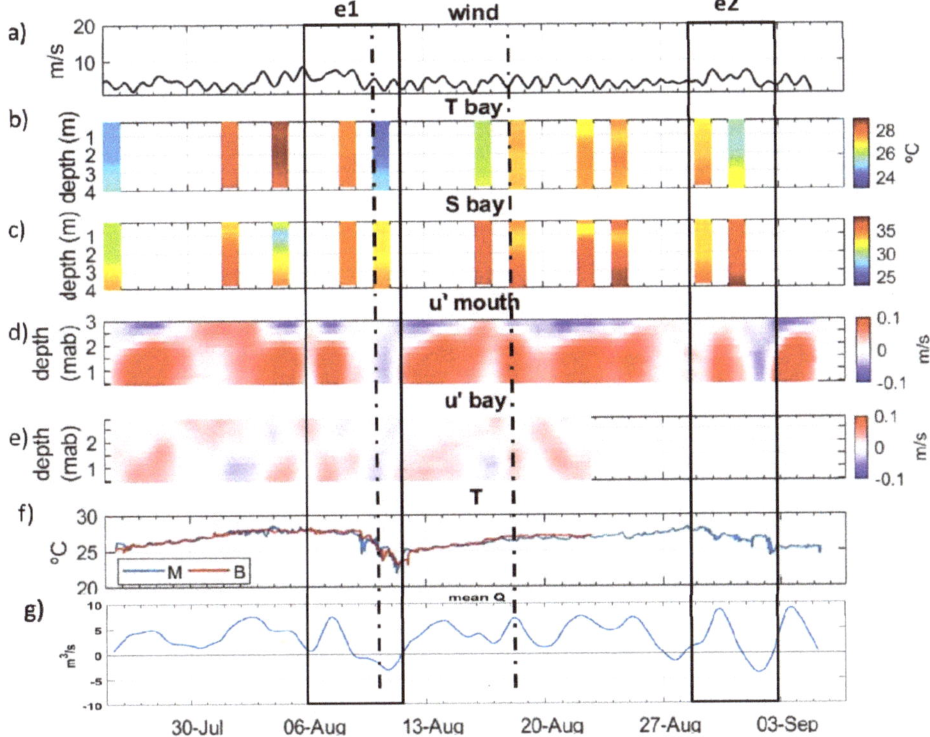

**Figure 8.** From top to bottom, wind speed (**a**), time series of temperature at a central point of the bay (**b**), time series of salinity at a central point of the bay (**c**), alongshore currents in mouth (**d**), alongshore currents in the bay (**e**), and bottom temperature (**f**) in mouth (blue) and bay (red), mean flow (positive inflow) in the mouth (**g**). The black boxes show the NW events. The dotted lines mark the days of the vertical sections shown in Figure 10.

The water temperature observations during FANGAR-I exhibited variations of between 23 °C and 29 °C (Figure 8b and f). The maximum temperature was reached on 4 August after the ambient temperatures reached 30 °C (Figure 8). However, just two days later on 6 August, the water temperature dropped drastically to 23 °C, indicating a difference of 6 °C. This event coincided with the presence of the NW winds, as can be seen in Figure 8. Once the wind stopped, the temperatures rapidly increased again, reaching 28 °C, as can be seen in Figure 8b–f. As noted in the previous paragraph, during episodes E1 and E2, the wind events started on 6 and 27 August, respectively. For both episodes, the temperatures began to drop from those dates and persisted during the subsequent days when the wind died down (Figure 8b–f). This temperature decrease was associated with a cold NW wind, alongside the likely contribution of an open sea entrance (i.e., colder) through the mouth, indicated by the reduction in water temperature observed a few days later, when the wind stopped.

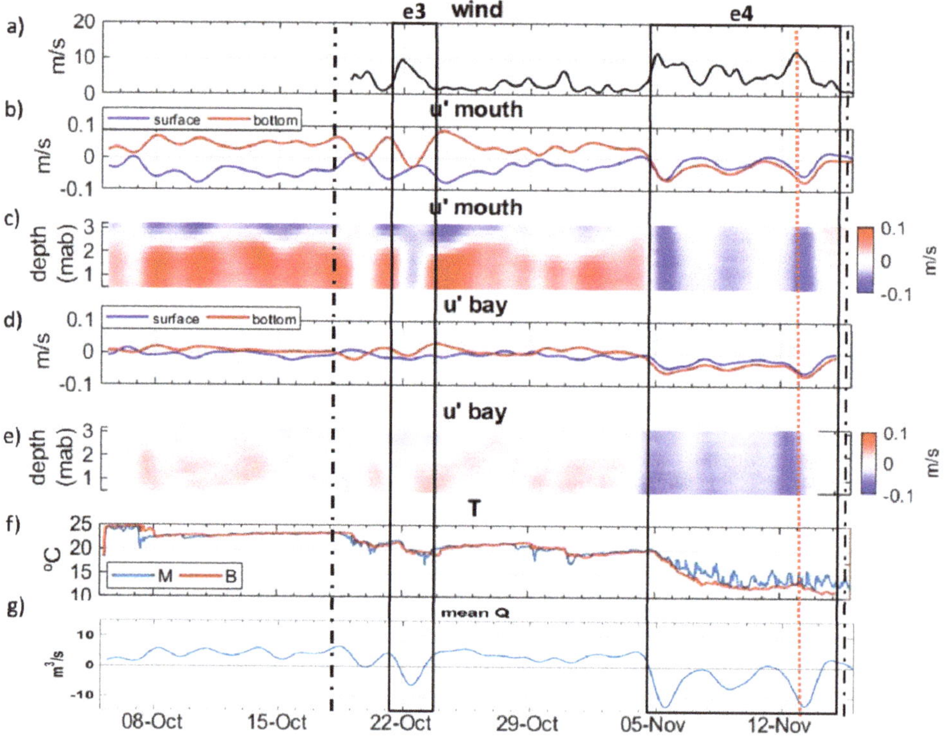


**Figure 9.** From top to bottom, wind speed (**a**), alongshore currents in the mouth (**b**), Hovmöller diagram in the mouth for the alongshore currents (**c**), alongshore currents in the bay (**d**) and Hovmöller diagram in the bay for the alongshore currents (**e**), and bottom temperature (**f**) in the mouth (blue) and the bay (red), mean flow (positive inflow) in the mouth (**g**). The black boxes show the NW events. The dotted lines mark the days of the vertical sections shown in Figure 10.

During FANGAR-II, the water temperature was lower than FANGAR-I because the ambient temperature decreased (i.e., autumn season) (Figure 9a). Throughout FANGAR-II, water temperatures between 11 °C and 25 °C were recorded. The campaign began with water temperatures of 22 °C, until 19 October saw a small decline in the sea bottom water temperature (20 °C) (Figure 9f). This could be associated with E3, when another drop in the water temperature occurred, falling to 18 °C. Once the NW wind stopped, the water temperatures rose slightly again to 20 °C, until 5 November when there was another episode of NW winds (E4). Note that before this event the water temperatures in the bay and in the mouth were similar, whereas from 5 November there were oscillations of about 4 °C in the data from the mouth, as can be seen in Figure 9f. This is due the diurnal cycle of heat fluxes (day/night) during those days.

Figure 10 shows a vertical section of the bay area (A–A′, Figure 1) for four different situations. Figure 10a,b depicts a situation prior to NW winds during FANGAR-I and during FANGAR-II, respectively. In these cases, a low vertical gradient in temperature could be observed (0.5 °C). Focusing on salinity, there was an evident gradient, with less salty surface waters (freshwater contributions, 31–33 for Figure 10a, 32–34 for Figure 10b) and deeper, saltier waters characteristic of ocean waters (37 for Figure 10a, 38 for Figure 10b). Figure 10c,d shows the T/S values during or immediately after NW episodes (i.e., E1 and E4). There was a clear horizontal gradient for both temperature and salinity in Figure 10c (i.e., a decreasing gradient as it enters the inner part of the bay), while in Figure 10d the salinity was homogenized in the water column and the temperature also decreased towards the inner part of the bay.

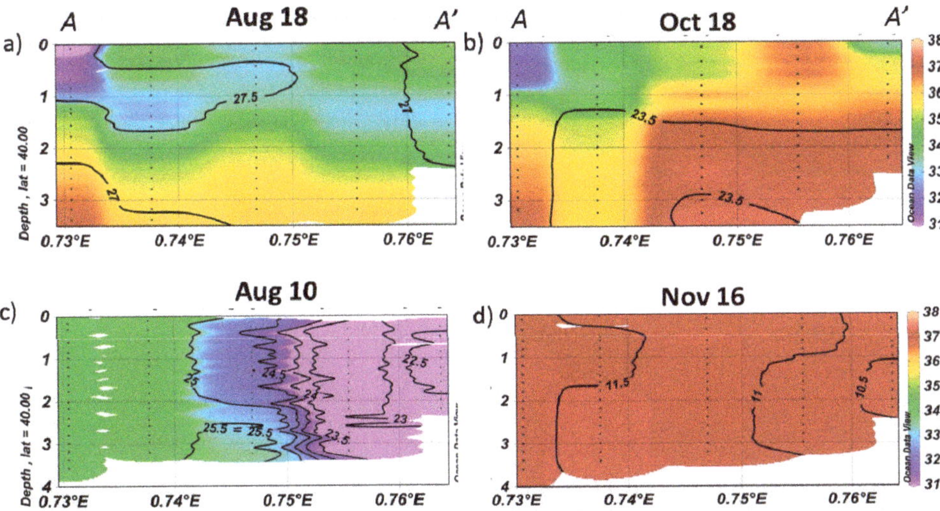


**Figure 10.** Vertical transect of (**a**) stratification summer situation, (**b**) stratification autumn situation, (**c**) situation during episode E1, (**d**) situation during episode E4. The color bar shows salinity values; black lines show temperature values.

### 3.3. Numerical Experiments

Figure 11 shows the results of the numerical model in terms of surface and bottom water currents, with no more variables than the force of the wind pushing the water. SIMU 1 (up-bay wind) indicated an expected water circulation pattern for the homogeneous flow (surface water following the wind direction and return flow at the bottom). However, when a varying bathymetry was considered (SIMU 2), the surface flow in the central axis reduced significantly in comparison to the shallower shore where the bottom friction is very relevant due to its shallowness. Tests have been made with different bottom roughness and it was found that the magnitude varies but the circulation pattern does not. In this case, an outward flow was observed in the central axis of the estuary in the bottom layers, opposite to the direction of the wind, as noted in the observations of the campaigns in the previous section. A homogeneous upward flow is observed at the shore. This pattern was reproduced for upward and smaller winds (SIMU 3). For the case down-bay wind, the picture differed significantly: downward flow towards the shore and upward flow in the axis and the bottom layers (SIMU 4). Therefore, the numerical experiments revealed transversal variability in flow direction due to the bathymetry. SIMU 5, 6 and 7 reproduced the previous cases but a stretching at the bay entrance is added. In this case, the pictures do not significantly differ in relation of the upward/downward flow in the shore/central axis for bottom layers in relation to the open mouth. The Coriolis effect was negligible in idealized cases when comparing SIMU 1 and SIMU 8. Finally, sensitivity test of the bottom friction parametrization has not shown relevant influence on the water circulation patterns affecting mainly the water speed.

Focussing on the impact of the wind on water circulation in the bay entrance, the transversal barotropic velocities are shown in Figure 12. This figure shows how the laterally variable bathymetry induces axially symmetric transverse structure (see inshore velocities near the boundaries and offshore in the central section of the bay entrance for SIMU 2; note that barotropic velocities for SIMU 1 its near to zero). Opposite water circulation pattern occurs for downwind simulation (SIMU 4). Bay mouth stretching (SIMU 6) has a similar pattern than SIMU 4 but increasing the spatial gradients between inflow and outflow (not shown). In this case, the magnitude of the transversal barotropic velocity is lower, reducing also the flow exchange expected between the bay and the open sea if the width mouth is reduced.

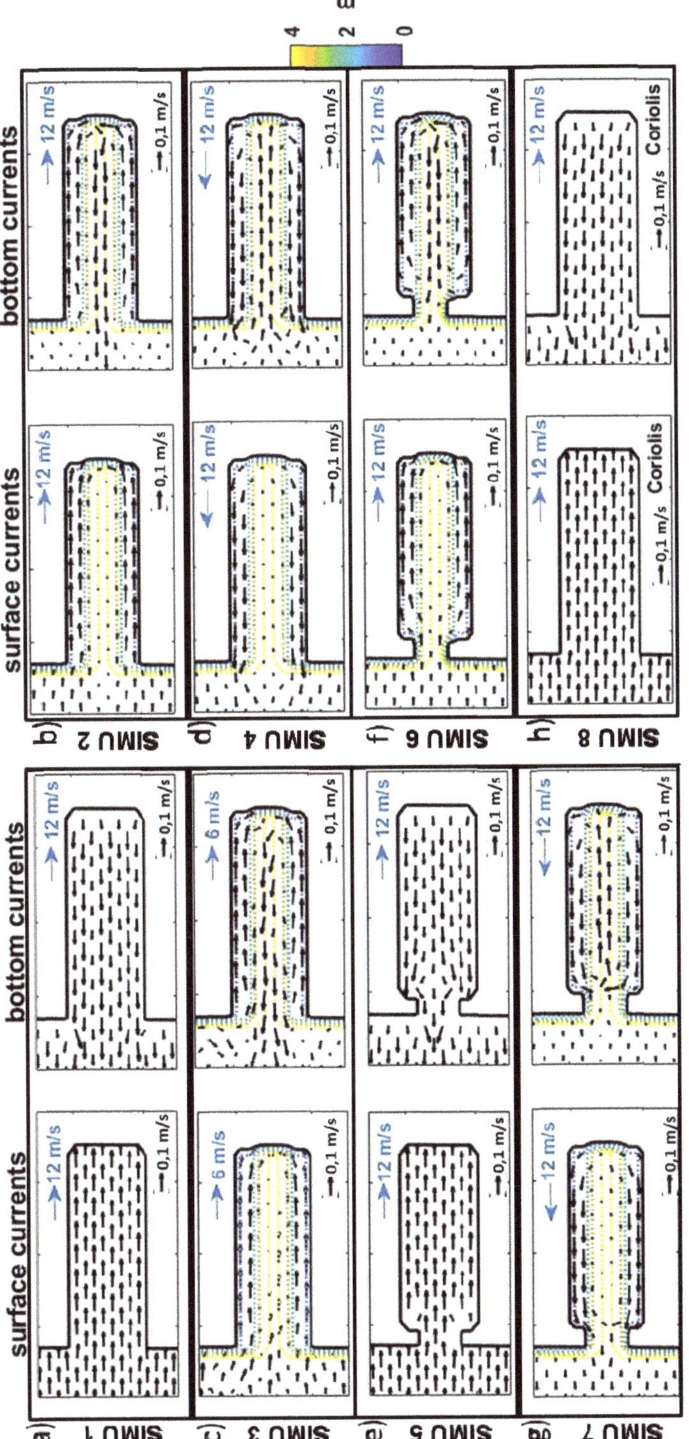

**Figure 11.** Representation of the simulations: (**a**) SIMU 1, surface and bottom; (**b**) SIMU 2, surface and bottom; (**c**) SIMU 3, surface and bottom; (**d**) SIMU 4, surface and bottom; (**e**) SIMU 5, surface and bottom; (**f**) SIMU 6, surface and bottom; (**g**) SIMU 7, surface and bottom; (**h**) SIMU 8, surface and bottom. Blue arrows show the wind direction with the intensity wind and black arrows show the water current intensity.

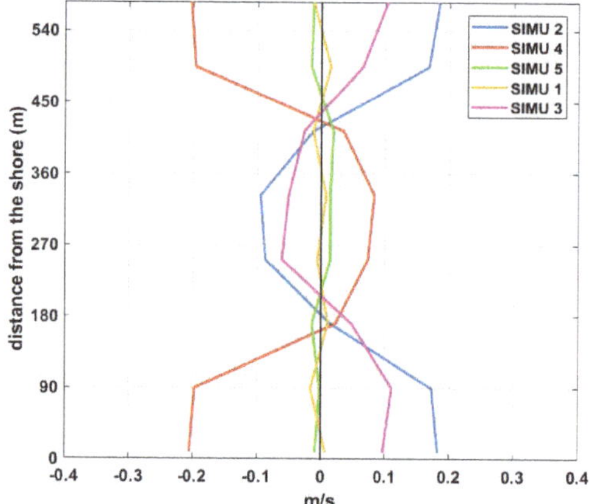

**Figure 12.** Barotropic velocities (negative outflow) in the entrance section for selected simulations.

## 4. Discussion

Local winds have the predominant influence on water circulation in small, shallow and micro-tidal estuaries and coastal bays. Such influence may lead to complex water circulation patterns due to bathymetric and geometric effects. For instance, Geyer [5] has found that geometric constrictions restrict wind-induced circulation, resulting in strong fronts between well-mixed reaches in two small Cape Cod (USA) estuaries. Scully et al. [32] proposed appeared during NW strong wind periods: up-estuary wind weakens and even reverses the shear and reduces the stratification. Li et al. [33] suggested that wind mixing may dominate over wind straining under typical wind forcing conditions in Chesapeake Bay. More recently, Xie et al. [34] tried to explain the apparent contradiction between these two studies and revealed opposite responses of the upper and lower bay regions to axial winds. Furthermore, Coogan et al. [35] have identified substantial transverse variability in Mobile Bay (USA), including the influence of the bathymetric effects previously postulated by [9]. Shallow depths accentuate the wind's influence, sometimes leading to complex water circulation according to observational investigations where the considerable variability of time series is observed [5,7,34]. Llebot et al. [6] have suggested that the origin of high variability of water currents corresponds to wind variability which is more effective in shallow depths in comparison to deep waters. For instance, the sea breeze represents a sequence covering a wide range of wind directions and intensities. Fangar Bay provides a good example of such estuaries (i.e. shallow and micro-tidal), where the bathymetry and the geometry are relevant, as well as being a small-scale bay, which accentuates the action of the wind. The data analysis conducted here, based on two intensive field campaigns, revealed a differentiated pattern in terms of the function of the location (mouth or inner part of the estuary) and the wind conditions (calm or intense). During calm conditions, estuarine circulation linked with strongly stratified conditions were identified in the mouth, associated with strongly stratified conditions. In the inner part, stratification in the water column also occurred, but water circulation did not show a clear pattern. Mixing in the water column occurs during intense wind with substantial modifications in temperature and salinity profiles as will be analyzed later. Particularly during up-bay wind episodes, outward flow conditions were observed in the mouth.

The numerical experiments revealed an axially symmetric transverse structure due to laterally varying bathymetry: up-bay flow occurs in the lateral shoals and down-bay flow in the central channel for up-estuary wind pulses. Consequently, the numerical model's results supported the observed divergence between wind and water flow direction in the center of the mouth in Fangar Bay. This combination of observations and a numerical model confirm that there is a strong transverse variability on water flow when bathymetric effects are considered. Comparing the SIMU 1 and SIMU 2 numerical experiments provides an illustrative example of the influence of bathymetry (see Figure 11). These results are consistent with Alekseenko et al. [36] for a coastal lagoon in the Mediterranean Sea (opposite flow in comparison to wind direction in a central part of the lagoon). Moreover, Sanay and Valle-Levinson [10] and Narváez et al. [3] found similar results for large bays than the one cited in this paper, highlighting the value of numerical experiments seeking the fundamental role played by bathymetry in estuarine circulation. Complementing these contributions, we also focused on geometric stretching in the mouth as well as up-bay wind effects using a numerical model in idealized conditions. The geometric stretching appeared to have a limited effect on water circulation, modulating the water exchange in the mouth (see SIMU 6 vs. SIMU 2). This reduction in the mouth width accords with [5], who has suggested that constrictions block wind-induced flushing and affect along-estuary salinity gradients. The mixing mechanism in very-shallow bays due strong winds has been illustrated by [6] in a similar domain than Fangar Bay. During these episodes pycnocline is tilted and becomes nearly vertical. In these cases, mixing is fast and largely driven by shear, and the interface becomes less defined. Numerical experiments based on our simple geometric model, including density variability, are suggested for further experiments focusing on the relationship between bathymetry variability and the evolution of the hydrographic structure. Remote winds may also produce water currents which flow against the local wind within the estuary. This behavior has been observed in Delaware or Chesapeake estuaries for low frequencies and is caused by a coastal Ekman set up due to remote wind effect [4,37]. The remote wind effect seems not relevant due to the small scale and depth of the Fangar Bay, being the local frictional effects prevailing such as will be discussed in the next paragraph. Additionally, energetic wind events in Ebro delta (i.e., north-easterlies and north-westerlies winds) tends to affect a substantial percentage of the continental shelf being the wind spatial variability small [38]. However, the implementation of a nested numerical model using real configuration may solve the remote wind effect in Fangar Bay.

Observations in Fangar Bay indicated that the water circulation was complex during calm periods: current velocities were very small and lacked a clear pattern (see Figures 5, 8 and 9), consistent with other contributions mentioned above. Momentum balance can facilitate analysis of the mechanism governing water circulation [38]. In very shallow domains, the bottom frictional term ($\tau^B$) in depth-averaged momentum balance equations (i.e., $\tau^B/\rho H$) may prove relevant (i.e., $H^{-1}$ relation), indicating a substantial effect on water fluctuations. This means that local bathymetric disturbances may have a relevant role in modifying the flow. Similar findings were attained by Noble et al. [9], observing how a complex, non-linear residual force leads to a stratified estuary in the case of Mobile (USA). These studies reveal that during calm or weak wind periods, being shallow and narrow areas, estuarine circulation is weak and occurs at the expense of frictional parameters, rendering a stable circulation pattern non-existent.

Although water circulation in the inner Fangar Bay proved chaotic and complex, periods of wind intensity of up 9 m·s$^{-1}$ substantially modified water circulation within the Bay's basin. The effects of the energetic NW winds (up-bay wind) on the hydrodynamics could be clearly observed in the results. The observational results in Fangar Bay revealed that mixing occurred when the wind was equal to 10 m·s$^{-1}$ (Beaufort scale 5), as seen in Figure 10c,d. Weaver et al. [39] similarly observed that only a constant high-magnitude wind was capable of flushing the Indian River Lagoon, while Coogan et al. [35] noted that in Mobile Bay, winds of 10–15 m·s$^{-1}$ could break the strong stratification and completely mix the water column until strong river discharges re-stratified the water again. Furthermore, Geyer [5] obtained data from two estuaries on Cape Cod (USA) where weak circulation during onshore winds was altered by wind force, inhibiting circulation in the estuary. Mixing was even observed during summer, when temperatures reached 28 °C (Figure 8f). Persistent warm conditions in Fangar Bay have been associated with increases in mussel mortality and oyster growth [40]. The fact that the NW winds cool the water, as observed in the Figures 8 and 9, in a relatively short time aid aquaculture activity. Increased warm periods in recent years represent a significant economic threat to both aquaculture producers in the bay and authorities in the area.

Wind forcing in shallow waters generates stress and introduces turbulence directly to the water column. The data analysis in Fangar Bay revealed that when the rice channels were opened during calm periods, freshwater inflow predominated and stratification was observed in the water column. Homogenization of the water column occurred during strong NW winds (up-bay wind) episodes. The same was true in periods of closed channels, when stratification was less intense. The large-scale response to wind forced may be characterized using the Wedderburn number [6,41], estimated according to the stabilization effect of the stratification and the destabilization effect of the wind:

$$W = (\Delta \rho g H^2)/L\tau_W \tag{1}$$

where $\tau_w$ = wind stress along the channel, $L$ = length of the channel, $\Delta \rho$ = density change over $L$, $H$ = channel depth and $g$ = gravitational acceleration. For a wind of 12 m·s$^{-1}$, W was equal to 0.68 (i.e., W < 1), so the wind produced a rapid, shear-driven mixture as the pycnocline tilted to become almost vertical, with the consequent development of horizontal density gradients (see Figure 10). In the case of weak winds (6 m·s$^{-1}$), the result was W equal to 2.75 (W >> 1), so in this case the pycnocline deepened slowly due to stirring. In the observations presented here, it was seen that the episodes of wind at Beaufort scale 5 came from the NW with intensities of between 10–12 m·s$^{-1}$, leading to mixing. However, being a small-scale and shallow bay, winds greater than 9 m·s$^{-1}$ may have been able to homogenize the water column. This could clearly be observed during episode E1, when winds rapidly passed from 5 to 10 m·s$^{-1}$ and were able to stimulate mixing. When there were breezes (weaker winds, ≤6 m·s$^{-1}$), the bay remained stratified. These results agreed with the calculations of the Wedderdurn number, confirming that a moderately strong wind produces bay-wide mixing, due to the shallowness of the basin (4 m). For deeper depths (for example 10 m), the wind necessary to stimulate shear-driven mixing would be about 25 m·s$^{-1}$ (i.e., a wind of level 10 on the Beaufort scale). In larger bays, it is not possible for the entire water column to be mixed due to the greater role played by earth rotation and strong density gradients [3,10].

The Wedderburn number assumes that the wind is constant and blows from a longer cut-off ($T_c$). $T_c$ was the length of time needed for the effects of the boundaries to propagate to the open end of the estuary [11]:

$$T_c = \frac{L}{\sqrt{g'h}} \tag{2}$$

where L was the length of the basin, $g' = g\Delta\rho/\rho_0$ was the reduced gravity, $\Delta\rho$ represented the top-bottom density difference, $\rho_0$ was a reference water density and $h$ was the average depth. The relaxation times of the horizontal density gradients (cut-off times) for density differences of ~3 kg·m$^{-3}$, whose average depth was 2 m, were about 7 h. This means that a wind blowing during this time causes bay-wide water mixing. Therefore, in Fangar Bay, winds from the NW that usually blow for more than one day will cause mixing. The sea breeze, which is of lower intensity and blows for a shorter period than the cut-off period (Tc), is not capable of causing this mixing, as can be seen in Figures 8 and 9. After the wind ceases, the system reverts to its original state, with horizontal isopycnals. The time scale necessary to bring the system back to its original state could be calculated as [11]:

$$T_r = UL/g'H \tag{3}$$

where $H$ was the depth of the estuary and $U$ was the longitudinal velocity generated during the relaxation process, calculated in turn as:

$$U = \sqrt{g'H} \tag{4}$$

Given that $L$ was small, the order of magnitude was of just a few hours. Specifically, for Fangar Bay, $T_r$ ranges of five to nine hours implied very short relaxation periods in comparison to larger domains. A visual inspection suggested a time lag between wind decay and the apparition of estuarine circulation in Fangar Bay of this order of magnitude (see Figure 8).

The water renewal has a relevant influence on water quality within the bay. A simple assessment of the residence time may be computed dividing the volume of water by the flow rate that comes out of the bay (calculated from the velocities in Figure 12). In these cases, we obtain residence times of 4 d for SIMU 2 and 4, and 1 day for SIMU 1 and 5 and respectively. The rest of the simulations are in the same order of magnitude. From the observed change flow (Figures 8 and 9) the average residence time is estimated in 4 d. The study carried out by F-Pedrera Balsells et al. [42] compute the residence times in the bay in more detail and some possible modifications to help in the bay management.

Empirical orthogonal functions (EOF) estimated for the alongshore water current directions from the entire vertical profile were used to identify the modal variability in the water current measurements (Figure 13a–d). EOFs allow multivariable data to be decomposed into its main components. Along-axis EOF analysis depicted the prevalence of the first component in both periods: 58/76% (Mouth/Bay) in FANGAR-I and 80/90% (Mouth/Bay) in FANGAR-II. During weak wind or calm periods (i.e. most of the time in FANGAR-I), the main components of the EOF analysis showed crossing-zero (suggesting stratification) with baroclinic conditions (see Figure 13a–c). However, during periods of intense winds, as was true of FANGAR-II, the first component revealed a mixing situation, seen especially at the inner bay B point, with barotropic flows.

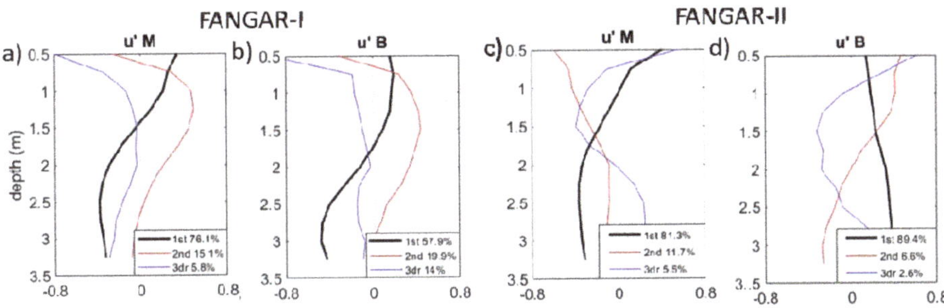

**Figure 13.** Empirical orthogonal functions (EOF) analysis for low-frequency filtered data during FANGAR-I (**a,b**) and FANGAR-II (**c,d**) along-shore currents. The legend shows the corresponding percentage of the explained variability.

Rotational effects in small bays and estuaries were expected to be small. Using the geometry of Fangar Bay, the Kelvin number was estimated, which considers the characteristic width of the estuary and the internal radius of deformation [43]:

$$K_e = B/R_{in} \tag{5}$$

where $R_{in}$ was equal to $(g'H)^{(1/2)}/f$, $g'$ = reduced gravity, $H$ = channel depth and $f$ = Coriolis force. Rotational effects play a secondary role on the basin circulation when $K^e < 1$, and they can be considered negligible when $K^e \ll 1$. In the case of Fangar Bay, which is only 2 km wide and 4 m deep, $K^e = 0.39$ ($\ll 1$) suggesting that the Coriolis effect is of little importance in terms of water circulation. This was supported by a comparison of SIMU 1 and SIMU 8 (Figure 11a,h), where the differences in water currents were negligible.

Questions remain open from observational analysis, in particular the hydrodynamic response during downwind and the spatial variability of the currents. From a numerical point of view, numerical modeling effort may include the bottom roughness sensitivity test, longitudinal and vertical density variations, sea-level oscillations, effect density gradients and baroclinic process due to river discharge sources. The numerical model implemented here is focused on solving the bathymetric effect which affect the wind-driven water circulation due to its shallowness under idealized perspective. So, realistic model implementation is a challenge to evaluate some remaining questions addressed before.

## 5. Conclusions

The dynamics in small-scale, shallow and micro-tidal estuaries were found to be characterized by very complex circulation in the inner part because of the effectiveness of local wind forcing, as well as a tendency towards stratification due to freshwater inputs that break down in episodes of intense wind lasting a few hours (O~$10^1$ h). Once the winds stop blowing, the system returns to its original state with O~$10^1$ h, so they are systems with very short relaxation periods. During these episodes of intense winds, bathymetry effects may produce an axially symmetrical transverse structure with outflow on the axis of the central channel, opposite to the wind direction, alongside inflow in shallow lateral areas. Furthermore, being a small-scale bay (2 km wide) and very shallow (average depth 2 m) these winds are capable of mixing all the water inside the bay, and being a narrow domain, the rotation effect is negligible. According to the data obtained from field campaigns and numerical idealized experiments in Fangar Bay, we suggest that a small-scale and shallow bay can be defined as being of 20 km$^2$ area and about 4 m depth, respectively.

**Author Contributions:** Conceptualization, methodology, M.F-P.B., M.G. and M.E.; software, M.F-P.B., M.G. and M.E.; validation, M.G. and M.E.; formal analysis, M.F-P.B.; investigation, M.F-P.B., P.C., M.G. and M.E.; resources, M.G. and M.E.; data curation, M.F-P.B.; writing—original draft preparation, M.F-P.B.; writing—review and editing, M.G., M.E. and A.S.-A.; visualization, M.G.; supervision, M.G., M.E. and A.S.-A.; project administration, M.E. and A.S.-A.; funding acquisition, M.E. and A.S.-A. All authors have read and agreed to the published version of the manuscript.

**Funding:** This research received funding from the Ecosistema-BC CTM2017-84275-R AEI/FEDER, UE.

**Acknowledgments:** We would like to thank Jordi Cateura and Joaquim Sospedra (LIM-UPC, Barcelona, Spain) and Margarita Fernandez (IRTA, Sant Carles de la Ràpita, Spain) for the data acquisition campaign and Alfredo Aretxabaleta (USGS, Woods Hole, USA) for his suggestion in previous versions of the manuscript. Also by the Government of Catalonia, in the framework of the FEMP-FANGAR2017 project and by the Direcció General de Pesca of the Government of Catalonia (DGP). Data from that project used in the current paper are archived in a public repository accessible at https://doi.org/10.5281/zenodo.3756624. As a group, we would like to thank the Secretaria d'Universitats i Recerca del Departament d'Economia i Coneixement de la Generalitat de Catalunya (2017SGR773).



## References

1. Geyer, W.R.; MacCready, P. The estuarine circulation. *Annu. Rev. Fluid Mech.* **2014**, *46*, 175–197. [CrossRef]
2. Csanady, G.T. Wind-Induced Barotropic Motions in Long Lakes. *J. Phys. Oceanogr.* **1973**, *3*, 429–483. [CrossRef]
3. Narváez, D.A.; Valle-Levinson, A. Transverse structure of wind-driven flow at the entrance to an estuary: Nansemond River. *J. Geophys. Res. Ocean.* **2008**, *113*, 1–9. [CrossRef]
4. Wong, K.-C.; Valle-Levinson, A. On the relative importance of the remote and local wind effects on the subtidal exchange at the entrance to the Chesapeake Bay. *J. Mar. Res.* **2002**, *60*, 477–498. [CrossRef]
5. Geyer, W.R. Influence of wind on dynamics and flushing of shallow estuaries. *Estuar. Coast. Shelf Sci.* **1997**, *44*, 713–722. [CrossRef]
6. Llebot, C.; Rueda, F.J.; Solé, J.; Artigas, M.L.; Estrada, M.; Spitz, Y.H.; Solé, J.; Estrada, M. Hydrodynamic states in a wind-driven microtidal estuary (Alfacs Bay). *J. Sea Res.* **2014**, *85*, 263–276. [CrossRef]
7. Cerralbo, P.; Grifoll, M.; Espino, M. Hydrodynamic response in a microtidal and shallow bay under energetic wind and seiche episodes. *J. Mar. Syst.* **2015**, *149*, 1–13. [CrossRef]
8. Grifoll, M.; Cerralbo, P.; Guillén, J.; Espino, M.; Boye Hansen, L.; Sánchez-Arcilla, A. Characterization of bottom sediment resuspension events observed in a micro-tidal bay. *Ocean Sci.* **2019**, *15*, 307–319. [CrossRef]

9. Noble, M.A.; Schroeder, W.W., Jr.; Ryan, H.F.; Gelfenbaum, G. Subtidal circulation patterns in a shallow, highly stratified estuary: Mobile Bay, Alabama. *J. Geophys. Res. Ocean.* **1996**, *101*, 625–689. [CrossRef]

10. Sanay, R.; Valle-Levinson, A. Wind-Induced Circulation in Semienclosed Homogeneous, Rotating Basins. *J. Phys. Oceanogr.* **2005**, *35*, 2520–2531. [CrossRef]

11. Llebot, C. Interactions between Physical Forcing, Water Circulation and Phytoplankton Dynamics in a Microtidal Estuary. Ph.D. Thesis, Catalonia University of Technology, Barcelona, Spain, 2010.

12. Valle-Levinson, A.; Delgado, J.A.; Atkinson, L.P. Reversing water exchange patterns at the entrance to a semiarid coastal lagoon. *Estuar. Coast. Shelf Sci.* **2001**, *53*, 825–838. [CrossRef]

13. Cerralbo, P.; Grifoll, M.; Valle-Levinson, A.; Espino, M. Tidal transformation and resonance in a short, microtidal Mediterranean estuary (Alfacs Bay in Ebre delta). *Estuar. Coast. Shelf Sci.* **2014**, *145*, 57–68. [CrossRef]

14. Ochoa, V.; Riva, C.; Faria, M.; López, M.; Alda, D.; Barceló, D.; Fernandez, M.; Roque, A.; Barata, C. Science of the Total Environment Are pesticide residues associated to rice production affecting oyster production in Delta del Ebro, NE Spain ? *Sci. Total Environ.* **2012**, *437*, 209–218. [CrossRef]

15. Camp, J.; Delgado, M. Hidrogafia de las bahías del delta del Ebro. *Inv.Pesq.* **1987**, 351–369.

16. Carrasco, N.; Arzul, I.; Berthe, F.C.J.; Fernández-Tejedor, M.; Durfort, M.; Furones, M.D. Delta de l'Ebre is a natural bay model for Marteilia spp. (Paramyxea) dynamics and life-cycle studies. *Dis. Aquat. Organ.* **2008**, *79*, 65–73. [CrossRef] [PubMed]

17. Archetti, G.; Bernia, S.; Salvà-Catarineu, M. Análisis de los vectores ambientales que afectan la calidad del medio en la bahía del Fangar (Delta del Ebro) mediante herramientas SIG. *Rev. Int. Cienc. Tecnol. Inf. Geogr.* **2010**, *10*, 252–279.

18. Bolaños, R.; Jorda, G.; Cateura, J.; Lopez, J.; Puigdefabregas, J.; Gomez, J.; Espino, M. The XIOM: 20 years of a regional coastal observatory in the Spanish Catalan coast. *J. Mar. Syst.* **2009**, *77*, 237–260. [CrossRef]

19. Grifoll, M.; Navarro, J.; Pallares, E.; Ràfols, L.; Espino, M.; Palomares, A. Ocean-atmosphere-wave characterisation of a wind jet (Ebro shelf, nw mediterranean sea). *Nonlinear Process. Geophys.* **2016**, *23*, 143–158. [CrossRef]

20. Garcia, M.A.; Ballester, A.; Garcia, M.; Ballester, A. Notas acerca de la meteorología y la circulación local en la región del delta del Ebro. *Inv.Pesq.* **1984**, *48*, 469–493.

21. Muñoz, I. Limnología de la part baixa del riu ebre i els canals de reg: Els factors fisico-quimics, el fitoplancton i els macroinvertebrats bentonics. Ph.D. Thesis, Departamento de Ecología, Facultad, Facultad de Biología, Universidad de Barcelona, Barcelona, Spain, 1990.

22. Automatic Water Quality Information System, DEL EBRO, CHE-Hydrographic Confederation, Quality Alert Network, SAICA Project, 2013. Available online: https://www.saica.co.za/ (accessed on 30 January 2020).

23. Pérez, M.; Camp, J. Distribución espacial y biomasa de las fanerógamas marinas de las bahías del delta del Ebro. *Inv. Pesq.* **1986**, *50*, 519–530.

24. Shchepetkin, A.F.; McWilliams, J.C. The regional oceanic modeling system (ROMS): A split-explicit, free-surface, topography-following-coordinate oceanic model. *Ocean Model.* **2005**, *9*, 347–404. [CrossRef]

25. Haidvogel, D.B.; Arango, H.; Budgell, W.P.; Cornuelle, B.D.; Curchitser, E.; Shchepetkin, A.F.; Sherwood, C.R.; Signell, R.P.; Warner, J.C.; Wilkin, J. Ocean forecasting in terrain-following coordinates: Formulation and skill assessment of the Regional Ocean Modeling System. *J. Comput. Phys.* **2007**, 1–30. [CrossRef]

26. Cerralbo, P.; F.-Pedrera Balsells, M.; Grifoll, M.; Fernandez, M.; Espino, M.; Cerralbo, P.; Mestres, M.; Sanchez-Arcilla, A. Use of a hydrodynamic model for the management of the water renovation in a coastal system. *Ocean Sci. Discuss.* **2019**, *15*, 215–226. [CrossRef]

27. Cerralbo, P.; Grifoll, M.; Moré, J.; Sairouní Afif, A.; Espino, M.; Bravo, M. Wind variability in a coastal area (Alfacs Bay, Ebro River delta). *Adv. Sci. Res.* **2015**, *12*, 11–21. [CrossRef]

28. Carter, G.S.; Merrifield, M.A. Open boundary conditions for regional tidal simulations. *Ocean Model.* **2007**, *18*, 194–209. [CrossRef]

29. Warner, J.C.; Geyer, W.R.; Lerczak, J.A. Numerical modeling of an estuary: A comprehensive skill assessment. *J. Geophys. Res. Ocean.* **2005**. [CrossRef]

30. Niedda, M.; Greppi, M. Tidal, seiche and wind dynamics in a small lagoon in the Mediterranean Sea. *Estuar. Coast. Shelf Sci.* **2007**, *74*, 21–30. [CrossRef]

31. Duchon, C.E. Lanczos filtering in one and two dimensions. *J. Appl. Meteorol.* **1979**, *18*, 1016–1022. [CrossRef]

32. Scully, M.E.; Friedrichs, C.; Brubaker, J. Control of estuarine stratification and mixing by wind-induced straining of the estuarine density field. *Estuaries* **2005**, *28*, 321–326. [CrossRef]
33. Li, Y.; Li, M. Effects of winds on stratification and circulation in a partially mixed estuary. *J. Geophys. Res. Ocean.* **2011**, *116*, C12012. [CrossRef]
34. Xie, X.; Li, M. Effects of Wind Straining on Estuarine Stratification: A Combined Observational and Modeling Study. *J. Geophys. Res. Ocean.* **2018**, *123*, 2363–2380. [CrossRef]
35. Coogan, J.; Dzwonkowski, B.; Park, K.; Webb, B. Observations of Restratification after a Wind Mixing Event in a Shallow Highly Stratified Estuary. *Estuaries Coasts* **2019**, *43*, 272–285. [CrossRef]
36. Alekseenko, E.; Roux, B.; Sukhinov, A.; Kotarba, R.; Fougere, D. Nonlinear hydrodynamics in a mediterranean lagoon. *Nonlinear Process. Geophys.* **2013**, *20*, 189–198. [CrossRef]
37. Garvine, R.W. A simple model of estuarine subtidal fluctuations forced by local and remote wind stress. *J. Geophys. Res.* **1985**, *90*, 11945. [CrossRef]
38. Ràfols, L.; Grifoll, M.; Jordà, G.; Espino, M.; Sairouní, A.; Bravo, M. Shelf Circulation Induced by an Orographic Wind Jet. *J. Geophys. Res. Ocean.* **2017**, *122*, 8225–8245. [CrossRef]
39. Weaver, R.J.; Johnson, J.E.; Ridler, M. Wind-Driven Circulation in a Shallow Microtidal Estuary: The Indian River Lagoon. *J. Coast. Res.* **2016**, *322*, 1333–1343. [CrossRef]
40. Ramón, M.; Fernández, M.; Galimany, E. Development of mussel (Mytilus galloprovincialis) seed from two different origins in a semi-enclosed Mediterranean Bay (N.E. Spain). *Aquaculture* **2007**, *264*, 148–159. [CrossRef]
41. Shintani, T.; De La Fuente, A.; Niño, Y.; Imberger, J. Generalizations of the wedderburn number: Parameterizing upwelling in stratified lakes. *Limnol. Oceanogr.* **2010**, *55*, 1377–1389. [CrossRef]
42. F-Pedrera Balsells, M.; Mestres, M.; Fernández, M.; Cerralbo, P.; Espino, M.; Grifoll, M.; Sánchez-Arcilla, A. Assessing Nature Based Solutions for Managing Coastal Bays. *J. Coast. Res.* **2020**, *95*, 1083–1087. [CrossRef]
43. Bolaños, R.; Brown, J.M.; Amoudry, L.O.; Souza, A.J. Tidal, Riverine, and Wind Influences on the Circulation of a Macrotidal Estuary. *J. Phys. Oceanogr.* **2013**, *43*, 29–50. [CrossRef]

MDPI

St. Alban-Anlage 66

4052 Basel

Switzerland

Tel. +41 61 683 77 34

Fax +41 61 302 89 18

www.mdpi.com

*Applied Sciences* Editorial Office

E-mail: applsci@mdpi.com

www.mdpi.com/journal/applsci